PERGAMON INTERNATIONAL LIBRARY
of Science, Technology, Engineering and Social Studies

The 1000-volume original paperback library in aid of education,
industrial training and the enjoyment of leisure

Publisher: Robert Maxwell, M.C.

Arithmetic Applied Mathematics

THE PERGAMON TEXTBOOK
INSPECTION COPY SERVICE

An inspection copy of any book published in the Pergamon International Library will
gladly be sent to academic staff without obligation for their consideration for course
adoption or recommendation. Copies may be retained for a period of 60 days from
receipt and returned if not suitable. When a particular title is adopted or recommended
for adoption for class use and the recommendation results in a sale of 12 or more copies,
the inspection copy may be retained with our compliments. The Publishers will be pleased
to receive suggestions for revised editions and new titles to be published in this important
International Library.

INTERNATIONAL SERIES IN

NONLINEAR MATHEMATICS: Theory, Methods and Applications
General Editors: V LAKSHMIKANTHAM and C P TSOKOS

VOLUME 1

Some other Pergamon Titles of Interest

BOWLER
Gravitation and Relativity

CONSTANTINESCU
Distributions and their Applications in Physics

GOODSTEIN
Fundamental Concepts of Mathematics, 2nd Edition

LAKSHMIKANTHAM & LEELA
An Introduction to Nonlinear Differential Equations in Abstract Spaces

LANDAU & LIFSHITZ
Course of Theoretical Physics (9 volumes)
A Shorter Course of Theoretical Physics (2 volumes)

MARTIN & HEWETT
Elements of Classical Physics

PATHRIA
The Theory of Relativity, 2nd Edition

PLUMPTON & TOMKYS
Theoretical Mechanics for Sixth Forms, 2nd Edition in SI Units

TAYLOR
Mechanics: Classical and Quantum

Important Research Journals*

Computers and Mathematics with Applications

Computers and Structures

Nonlinear Analysis: Theory, Methods and Applications

*Free specimen copies available on request

ARITHMETIC APPLIED MATHEMATICS

by

DONALD GREENSPAN

The University of Texas at Arlington, U.S.A.

PERGAMON PRESS

OXFORD · NEW YORK · TORONTO · SYDNEY · PARIS · FRANKFURT

U.K.	Pergamon Press Ltd., Headington Hill Hall, Oxford OX3 0BW, England
U.S.A.	Pergamon Press Inc., Maxwell House, Fairview Park, Elmsford, New York 10523, U.S.A.
CANADA	Pergamon of Canada, Suite 104, 150 Consumers Road, Willowdale, Ontario M2J 1P9, Canada
AUSTRALIA	Pergamon Press (Aust.) Pty. Ltd., P.O. Box 544, Potts Point, N.S.W. 2011, Australia
FRANCE	Pergamon Press SARL, 24 rue des Ecoles, 75240 Paris, Cedex 05, France
FEDERAL REPUBLIC OF GERMANY	Pergamon Press GmbH, 6242 Kronberg-Taunus, Hammerweg 6, Federal Republic of Germany

First edition 1980

British Library Cataloguing in Publication Data

Greenspan, Donald
Arithmetic applied mathematics. - (International
series in nonlinear mathematics; vol. 1)
1. Mathematical physics - Data processing
I. Title II. Series
530.1'5'02854044   QC20.7.E4   80-40295

ISBN 0-08-025047-5 (hardcover)
ISBN 0-08-025046-7 (flexicover)

In order to make this volume available as economically and as rapidly as possible the author's typescript has been reproduced in its original form. This method has its typographical limitations but it is hoped that they in no way distract the reader.

Printed in Great Britain by A. Wheaton & Co. Ltd, Exeter

6a6
2-6-82

Contents

Preface

In this book we will develop a computer, rather than a continuum, approach to the deterministic theories of particle mechanics. Thus, we will formulate and study new models of classical physical phenomena from both Newtonian and special relativistic mechanics by use only of arithmetic. At those points where Newton, Leibniz, and Einstein found it necessary to apply the analytical power of the calculus, we shall, instead, apply the computational power of modern digital computers. Most interestingly, our definitions of energy and momentum will be identical to those of continuum mechanics, and we will establish the very same laws of conservation and symmetry. The unifying concept will be that of the potential. In addition, the simplicity of our approach will yield simple models of complex physical phenomena and solvable dynamical equations for both linear and nonlinear behavior. The price we pay for such mathematical simplicity is that we must do our arithmetic at high speeds.

For their help in the preparation of this manuscript, I wish to thank John Collier, Robert LaBudde, and Judy Swenson. For their permission to quote freely from my monograph DISCRETE MODELS (Addison-Wesley, Reading, Mass., 1973), I wish to thank the editors of Addison-Wesley.

Donald Greenspan
Arlington, Texas

Chapter 1

Gravity

1.1 INTRODUCTION

One of the major functions of applied mathematics is the development and study of viable models of natural phenomena. Until recently, the concepts and methodology of the calculus were fundamental in such endeavors. However, with the development of modern digital computers, a new, additional type of modeling has emerged, called discrete modeling. In it, one uses the exceptional speed with which computers perform arithmetic, store and retrieve numbers, and execute basic logical decisions.

In this book we will explore discrete modeling as it relates to two of the major areas of applied mathematical research, namely, to Newtonian mechanics and to special relativistic mechanics. These two areas are the most substantial disciplines of deterministic physics. We will show how to formulate and develop each using only arithmetic and we will then explore the new types of models which emerge in a natural and simplistic way.

1.2 GRAVITY

It is always difficult to know how to begin correctly. In the physical sciences, one usually develops some intuition first by examining experimental results, either one's own or those of others. For this reason let us consider the following experiment with a force with which we are all aware, that is, gravity. If a particle P of mass m, situated at height h above ground, is dropped from a position of rest, one can measure its height x above ground every Δt seconds as it falls. For example, if one has a camera whose shutter time is Δt, then one can take a sequence of pictures at the times $t_k = k\Delta t$, $k = 0,1,2,\ldots$, and from the knowledge of h determine the heights $x_k = x(t_k)$ directly from the photographs by elementary ratio and proportion. Suppose, then, that this has been done, say, for $\Delta t = 1$ sec, and that, to the nearest foot, one finds

$$x_0 = 400, \quad x_1 = 384, \quad x_2 = 336, \quad x_3 = 256, \quad x_4 = 144, \quad x_5 = 0.$$

These data are recorded in column A of Table 1.1. By rewriting x_0, x_1, x_2, x_3, x_4 and x_5 as

$$x_0 = 400 - 0, \quad x_1 = 400 - 16, \quad x_2 = 400 - 64$$

$$x_3 = 400 - 144, \quad x_4 = 400 - 256, \quad x_5 = 400 - 400,$$

(which express the height above ground as the difference of the initial height and the distance fallen) and by factoring, one readily finds the interesting relationships

$$x_0 = 400 - 16(0)^2, \quad x_1 = 400 - 16(1)^2, \quad x_2 = 400 - 16(2)^2,$$

$$x_3 = 400 - 16(3)^2, \quad x_4 = 400 - 16(4)^2, \quad x_5 = 400 - 16(5)^2,$$

which can be written concisely as

$$x_k = 400 - 16(t_k)^2; \quad k = 0,1,2,3,4,5. \tag{1.1}$$

In the traditional manner, one would now interpolate from (1.1) to obtain the continuous formula

$$x = 400 - 16t^2, \quad 0 \le t \le 5, \tag{1.2}$$

from which, by differentiation, one would find

$$v(t) = x'(t) = -32t, \quad 0 \le t \le 5 \tag{1.3}$$

$$a(t) = v'(t) = -32, \quad 0 \le t \le 5. \tag{1.4}$$

The particle's velocities $v_0 = v(0)$, $v_1 = v(1)$, $v_2 = v(2)$, $v_3 = v(3)$, $v_4 = v(4)$, and $v_5 = v(5)$, at the times when the corresponding heights x_0, x_1, x_2, x_3, x_4, and x_5 have been recorded, are now determined directly from (1.3), and are recorded in column B of Table 1.1. The particle's accelerations at these times are determined from (1.4) and are recorded in column C of Table 1.1.

TABLE 1.1

Time	A Measured height	B Velocity by calculus	C Acceleration by calculus	D Velocity by arithmetic	E Acceleration by arithmetic
$t_0 = 0$	$x_0 = 400$	$v_0 = 0$	$a_0 = -32$	$v_0 = 0$	$a_0 = -32$
$t_1 = 1$	$x_1 = 384$	$v_1 = -32$	$a_1 = -32$	$v_1 = -32$	$a_1 = -32$
$t_2 = 2$	$x_2 = 336$	$v_2 = -64$	$a_2 = -32$	$v_2 = -64$	$a_2 = -32$
$t_3 = 3$	$x_3 = 256$	$v_3 = -96$	$a_3 = -32$	$v_3 = -96$	$a_3 = -32$
$t_4 = 4$	$x_4 = 144$	$v_4 = -128$	$a_4 = -32$	$v_4 = -128$	$a_4 = -32$
$t_5 = 5$	$x_5 = 0$	$v_5 = -160$	$a_5 = -32$	$v_5 = -160$	

Note that formulas (1.3) and (1.4), and the interesting conclusion that the acceleration due to gravity is constant, with the value -32, have all been <u>deduced</u> from the given distance measurements x_0, x_1, x_2, x_3, x_4, and x_5.

Let us show now that <u>all</u> the above conclusions could have been deduced without ever having introduced the concepts and methodology of the calculus. To do so, let us define the particle's velocity $v_k = v(t_k)$, $k = 0,1,2,3,4,5$, as an <u>average</u> (rather than <u>instantaneous</u>) rate of change of height with respect to time by the arithmetic formula

$$\frac{v_{k+1}+v_k}{2} = \frac{x_{k+1}-x_k}{\Delta t} \; ; \quad k = 0,1,2,3,4. \tag{1.5}$$

Since averaging procedures are both common and useful in the analysis of experimental data, the left-hand side of (1.5) is perfectly reasonable. Next, for computational convenience, let us rewrite (1.5) in the form

$$v_{k+1} = -v_k + 2(x_{k+1}-x_k)/(\Delta t); \quad k = 0,1,2,3,4. \tag{1.6}$$

Assuming that $v_0 = 0$ when a particle is dropped from a position of rest, one finds from (1.6) that

$$v_1 = -v_0 + 2(x_1-x_0)/(\Delta t) = 0 + 2(384-400)/1 = -32$$

$$v_2 = -v_1 + 2(x_2-x_1)/(\Delta t) = 32 + 2(336-384)/1 = -64$$

$$v_3 = -v_2 + 2(x_3-x_2)/(\Delta t) = 64 + 2(256-336)/1 = -96$$

$$v_4 = -v_3 + 2(x_4-x_3)/(\Delta t) = 96 + 2(144-256)/1 = -128$$

$$v_5 = -v_4 + 2(x_5-x_4)/(\Delta t) = 128 + 2(0-144)/1 = -160,$$

which are <u>identical</u> with the results of column B in Table 1.1, and are recorded in column D.

Next, since x_0 and v_0, but not a_0, are known initially, let us define a_k as the <u>average</u> (rather than <u>instantaneous</u>) rate of change of velocity with respect to time by the <u>arithmetic</u> formula

$$a_k = \frac{v_{k+1}-v_k}{\Delta t} \; ; \quad k = 0,1,2,3,4. \tag{1.7}$$

From the values v_k just generated, one finds from (1.7) that $a_0 = a_1 = a_2 = a_3 = a_4 = -32$, which are identical with entries in column C of Table 1.1, and are recorded in column E. Formula (1.7) does not allow a determination of a_5 because this would require knowing v_6. Nevertheless, the entries do indicate quite clearly that the acceleration due to gravity is constant, with the value -32.

Formulas (1.6) and (1.7) are both recursion formulas. Such formulas are solved numerically with exceptional speed on modern digital computers. Thus, even if the original distance measurements had been exceptionally voluminous, they could still have been recorded and analyzed quite easily.

Now, just because our arithmetic formulas (1.5) and (1.7) have given the same results as (1.3) and (1.4) does not mean that we have, as yet, a formulation which is of physical significance. Indeed, the physical significance of Newtonian mechanics is characterized by the laws of <u>conservation</u> of energy, linear momentum, and angular momentum, and by <u>symmetry</u>, that is, by the invariance of its laws of motion under fundamental coordinate transformations (see, e.g., reference [26]). Surprisingly enough, our approach to gravity will also yield conservation and symmetry. We will, however, confine attention at present only to the conservation of energy, not only for simplicity, but because of the intimate relationship between energy conservation and computational stability ([19], [81]).

For completeness, recall now the fundamental Newtonian dynamical equation:

$$F = ma,$$ (1.8)

the classical formula for kinetic energy K:

$$K = \frac{1}{2} mv^2,$$ (1.9)

and, for a falling body with $a = -32$, the formula for potential energy V:

$$V = 32mx.$$ (1.10)

The classical energy conservation law then states that if K_0 and V_0 are the kinetic and potential energies, respectively, at time $t_0 = 0$, while K_n and V_n are the kinetic and potential energies, respectively, at time $t_n > t_0$, then

$$K_n + V_n \equiv K_0 + V_0,$$ (1.11)

for all $t_n > t_0$.

It will be instructive, for the discussion later, to recall the derivation of (1.11). For this purpose, let P be at x_0 when $t = t_0$ and let P be at x_n when $t = t_n$. Then the work W done by gravity in the time interval $0 \leq t \leq t_n$ is defined by

$$W = \int_{x_0}^{x_n} F \, dx.$$ (1.12)

Hence,

$$W = \int_{x_0}^{x_n} m \, a \, dx = m \int_0^{t_n} a \, v \, dt = m \int_0^{t_n} \frac{d}{dt} \left(\frac{1}{2} v^2 \right) dt = \frac{1}{2} mv_n^2 - \frac{1}{2} mv_0^2,$$

so that

$$W \equiv K_n - K_0.$$ (1.13)

Note that (1.13) is independent of the actual structure of F. If one next recon-
siders (1.12) and uses the knowledge that F is gravity, then

$$W = -32m \int_{x_0}^{x_n} dx = -32mx_n + 32mx_0,$$

so that, from (1.10),

$$W \equiv -V_n + V_0. \tag{1.14}$$

Finally, conservation follows immediately from the elimination of W between
(1.13) and (1.14).

Let us now return to our arithmetic formulation. Recall that the experimental
data in column A of Table 1.1 were obtained from photographs at the distinct times
$t_k = k\Delta t$. For this reason, we will concentrate only on these times, so that (1.8)-
(1.10) need be considered only as follows:

$$F_k = ma_k; \quad k = 0,1,2,\ldots \tag{1.15}$$

$$K_k = \frac{1}{2}m(v_k)^2; \quad k = 0,1,2,\ldots \tag{1.16}$$

$$V_k = 32mx_k; \quad k = 0,1,2,\ldots . \tag{1.17}$$

In analogy with (1.12), define W_n, n = 1,2,3,... , by

$$W_n = \sum_{i=0}^{n-1} (x_{i+1}-x_i)F_i. \tag{1.18}$$

Then, by (1.5), (1.7) and (1.15)

$$W_n = m \sum_{i=0}^{n-1} (x_{i+1}-x_i)(\frac{v_{i+1}-v_i}{\Delta t})$$

$$= \frac{m}{2} \sum_{i=0}^{n-1} (v_{i+1}+v_i)(v_{i+1}-v_i) = \frac{m}{2}v_n^2 - \frac{m}{2}v_0^2 ,$$

so that

$$W_n \equiv K_n - K_0, \quad n = 1,2,3,\ldots \tag{1.19}$$

which is in complete analogy with (1.13) and is, also, independent of the structure
of F. On the other hand, since $a_k \equiv -32$, one has from (1.15) and (1.18) that

$$W_n = -32m \sum_{i=0}^{n-1} (x_{i+1}-x_i) = -32mx_n + 32mx_0,$$

so that, from (1.17)

$$W_n = -V_n + V_0, \quad n = 1,2,3,\ldots \tag{1.20}$$

in complete analogy with (1.14).

Finally, elimination of W_n between (1.19) and (1.20) yields

$$K_n + V_n \equiv K_0 + V_0, \quad n = 1,2,3,\ldots \tag{1.21}$$

in complete analogy with (1.11). Moreover, since K_0 and V_0 are determined from the initial conditions x_0 and v_0, it follows from (1.9), (1.10), (1.16) and (1.17) that $K_0 + V_0$ is the same in both (1.11) and (1.21), so that our strictly arithmetic approach conserves _exactly_ the same total energy, _independently_ of Δt, as does classical Newtonian theory.

It is also worth noting that in the derivations of (1.19) and (1.20), the telescopic sums

$$\sum_{i=0}^{n-1} (v_{i+1}^2 - v_i^2) \equiv v_n^2 - v_0^2$$

$$\sum_{i=1}^{n-1} (x_{i+1} - x_i) = x_n - x_0$$

play the same roles in the derivations of (1.19) and (1.20) as does integration in the derivations of (1.13) and (1.14).

Chapter 2

Long and Short Range Forces: Gravitation
and Molecular Attraction and Repulsion

2.1 INTRODUCTION

In classical physical modeling, the two basic types of forces which are important
are long range and short range forces ([1], [17], [26], [47], [49], [50], [69],
[77], [84], [99]). Typical of a long range force is gravitation, whose effect
throughout the solar system determines the natures of planets and moons. Typical
of a short range force is the local attractive and repulsive force between molecules, whose effects are fundamental, for example, in heat transfer, elasticity
and fluid flow phenomena. In this chapter we will examine both of these types of
forces in a fashion which continues and extends the development in Chapter 1.

2.2 GRAVITATION

In order to develop some intuition about gravitation, let us consider some new
possibilities for the force of gravity. Consider two bodies P_1 of mass m_1 and
P_2 of mass m_2, each on an X-axis, as shown in Fig. 2.1. Assume at first that
the mass of P_2 is almost negligible compared to that of P_1. Then, if P_1 were,
for example, the earth, and P_2 were a particle near the earth, we would expect
P_2 to fall to P_1 because of gravity. Now, suppose P_2 were a little more massive. It would probably still fall. Suppose, indeed, that P_2 were as massive as
P_1. Now it is not so clear as to what would happen. Moreover, if next we let P_2
be so massive that the mass of P_1 becomes the relatively negligible one, then,
indeed, we would even expect P_1 to fall to P_2. This suggests the possibility
that P_1 and P_2 might both have been in motion in all cases, but that only when
the mass of one particle was relatively negligible to the mass of the second did
the motion of the second become relatively negligible with respect to the motion of
the first. This behavior was assumed first by Isaac Newton and formalized in his
Law of Gravitation, which is given as follows.

Newton's Law of Gravitation

Each of two circular, homogeneous bodies, P_1 of mass m_1 and P_2 of mass m_2, exerts an attractive force on the other which is called the force of gravitation. The force which P_1 exerts on P_2 is equal in magnitude but opposite in direction to the force which P_2 exerts on P_1, and both forces act along the straight line joining the centers of P_1 and P_2. The magnitude $|F|$ of each of these forces is given by the $\frac{1}{r^2}$ law:

$$|F| = G\frac{m_1 m_2}{r^2},\tag{2.1}$$

where r is the distance between the centers of P_1 and P_2 and where G is a universal constant.

Various experiments, which date back from those of H. Cavendish in 1798 to those currently being performed by various space agency programs, have determined that, in cgs units,

$$G \sim (6.67)10^{-8}.\tag{2.2}$$

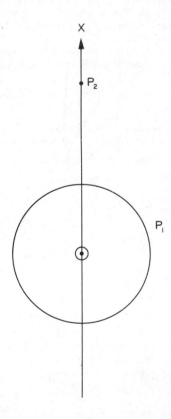

Fig. 2.1

With the intuition and ideas developed thus far, we are now ready to formulate a numerical model of Newtonian gravitation in more than one dimension. Of course, this is necessary if we wish to analyze, for example, the interesting motions of the planets.

2.3 BASIC PLANAR CONCEPTS

Let us show first how to generalize the basic concepts of position, velocity, and acceleration of Chapter 1. Since each of these quantities has both magnitude and direction, it is natural that we will use vector notation. Thus, the formulation, in essence, will include the three dimensional case, but, for simplicity, is developed as if one were interested at present only in the planar, or two dimensional, case.

For $\Delta t > 0$ and $t_k = k\Delta t$, $k = 0,1,2,\ldots$, let particle P of mass m be located at $\vec{r}_k = (x_k, y_k)$ at time t_k. If $\vec{v}_k = (v_{k,x}, v_{k,y})$ is the velocity of P at t_k, while $\vec{a}_k = (a_{k,x}, a_{k,y})$ is the acceleration of P at t_k, we will assume, in analogy with (1.5) and (1.7), that

$$\frac{\vec{v}_{k+1} + \vec{v}_k}{2} = \frac{\vec{r}_{k+1} - \vec{r}_k}{\Delta t} \,, \quad k = 0,1,2,\ldots, \tag{2.3}$$

$$\vec{a}_k = \frac{\vec{v}_{k+1} - \vec{v}_k}{\Delta t}, \quad k = 0,1,2,\ldots \,. \tag{2.4}$$

Of course, (2.3) and (2.4) are merely convenient and concise expressions for, respectively,

$$\frac{v_{k+1,x} + v_{k,x}}{2} = \frac{x_{k+1} - x_k}{\Delta t} \,,$$

$$\frac{v_{k+1,y} + v_{k,y}}{2} = \frac{y_{k+1} - y_k}{\Delta t} \,, \quad k = 0,1,2,\ldots, \tag{2.3'}$$

$$a_{k,x} = \frac{v_{k+1,x} - v_{k,x}}{\Delta t}$$

$$a_{k,y} = \frac{v_{k+1,y} - v_{k,y}}{\Delta t}, \quad k = 0,1,2,\ldots \,. \tag{2.4'}$$

To relate force and acceleration at each time t_k, we assume, in analogy with (1.15), that

$$\vec{F}_k = m\,\vec{a}_k, \tag{2.5}$$

where

$$\vec{F}_k = (F_{k,x}, F_{k,y}). \tag{2.6}$$

Of course, (2.5) is equivalent to

$$F_{k,x} = m\,a_{k,x}, \quad F_{k,y} = m\,a_{k,y}, \quad k = 0,1,2,\ldots \,. \tag{2.5'}$$

In analogy with (1.18), next define the work W_n by the inner product

$$W_n = \sum_{k=0}^{n-1} [(\vec{r}_{k+1} - \vec{r}_k) \cdot \vec{F}_k],$$
(2.7)

that is, by

$$W_n = \sum_{k=0}^{n-1} [(x_{k+1} - x_k) F_{k,x} + (y_{k+1} - y_k) F_{k,y}].$$
(2.8)

Now,

$$\sum_{k=0}^{n-1} (x_{k+1} - x_k) F_{k,x} = \sum_{0}^{n-1} (x_{k+1} - x_k) m\, a_{k,x}$$

$$= m \sum_{0}^{n-1} [(\frac{x_{k+1} - x_k}{t})(v_{k+1,x} - v_{k,x})]$$

$$= \frac{m}{2} \sum_{0}^{n-1} [(v_{k+1,x} + v_{k,x})(v_{k+1,x} - v_{k,x})]$$

$$= \frac{m}{2} \sum_{0}^{n-1} (v_{k+1,x}^2 - v_{k,x}^2)$$

$$= \frac{m}{2} v_{n,x}^2 - \frac{m}{2} v_{0,x}^2 .$$

Similarly,

$$\sum_{k=0}^{n-1} (y_{k+1} - y_k) F_{k,y} = \frac{m}{2} v_{n,y}^2 - \frac{m}{2} v_{0,y}^2 .$$

Thus, from (2.8),

$$W_n = \frac{m}{2} (v_{n,x}^2 + v_{n,y}^2) - \frac{m}{2} (v_{0,x}^2 + v_{0,y}^2).$$
(2.9)

We now define the kinetic energy K_i of P at time t_i by

$$K_i = \frac{m}{2} |\vec{v}_i|^2 = \frac{m}{2} (v_{i,x}^2 + v_{i,y}^2).$$
(2.10)

Then, from (2.9), one has

$$W_n = K_n - K_0,$$
(2.11)

in complete analogy with (1.19).

For convenience, we incorporate this into the following theorem.

Theorem 2.1. For $\Delta t > 0$, let $t_k = k\Delta t$. Let particle P of mass m be in motion in an XY plane. At time t_k, let P be located at $\vec{r}_k$, have velocity $\vec{v}_k$ and have acceleration $\vec{a}_k$. If W_n is defined by (2.7), then (2.11) is valid for all $n = 1,2,3,\ldots$.

2.4 DISCRETE GRAVITATION AND PLANETARY MOTION

For simplicity, let us begin by modeling a physical problem in which only one particle is in motion in the XY plane. A prototype problem of this type is the following, in which a planet, whose mass is relatively small compared to that of the sun, is in orbital motion around the sun.

Let the sun, whose mass is m_1, be positioned at the origin of the XY coordinate system. Let the position, velocity, and mass m_2 of a planet P be known at time t_0. Then, assuming that the sun's motion is negligible, we wish to determine the position (x_k,y_k) of P at each t_k, $k = 1,2,3,\ldots,n$, if the only acting force is gravitation. We define this gravitational force by the discrete formulas

$$F_k = (F_{k,x}, F_{k,y}) \tag{2.12}$$

$$F_{k,x} = - \frac{Gm_1m_2}{r_k r_{k+1}} \frac{\frac{x_{k+1}+x_k}{2}}{\frac{r_{k+1}+r_k}{2}} = - \frac{Gm_1m_2(x_{k+1}+x_k)}{r_k r_{k+1}(r_k+r_{k+1})} \tag{2.13}$$

$$F_{k,y} = - \frac{Gm_1m_2(y_{k+1}+y_k)}{r_k r_{k+1}(r_k+r_{k+1})} , \tag{2.14}$$

where G is the Newtonian constant (2.2), and where

$$r_k^2 = x_k^2 + y_k^2 . \tag{2.15}$$

Note immediately that (2.13) and (2.14) converge, in the limit as $\Delta t \to 0$, to the classical, conservative Newtonian formulas

$$F_x = - \frac{Gm_1m_2}{r^2} \cdot \frac{x}{r} ; \quad F_y = - \frac{Gm_1m_2}{r^2} \cdot \frac{y}{r} , \tag{2.16}$$

where

$$r^2 = x^2 + y^2 . \tag{2.17}$$

The significance of (2.13) and (2.14) over other possible formulas is that, independently of Δt, they conserve exactly the same energy as do their limiting, continuous forms, which is a consequence of the following discussion.

Consider, again, (2.7). Then (2.12)-(2.15) imply

$$W_n = \sum_{k=0}^{n-1} \left[(x_{k+1}-x_k) \left(- \frac{Gm_1 m_2 (x_{k+1}+x_k)}{r_k r_{k+1} (r_k+r_{k+1})} \right) \right.$$

$$\left. + (y_{k+1}-y_k) \left(- \frac{Gm_1 m_2 (y_{k+1}+y_k)}{r_k r_{k+1} (r_k+r_{k+1})} \right) \right]$$

$$= - Gm_1 m_2 \sum_0^{n-1} \left[\frac{x_{k+1}^2 - x_k^2 + y_{k+1}^2 - y_k^2}{r_k r_{k+1} (r_k+r_{k+1})} \right]$$

$$= - Gm_1 m_2 \sum_0^{n-1} \left[\frac{r_{k+1}^2 - r_k^2}{r_k r_{k+1} (r_k+r_{k+1})} \right]$$

$$= - Gm_1 m_2 \sum_0^{n-1} \left[\frac{r_{k+1} - r_k}{r_k r_{k+1}} \right]$$

$$= - Gm_1 m_2 \sum_0^{n-1} \left[\frac{1}{r_k} - \frac{1}{r_{k+1}} \right]$$

$$= \frac{-Gm_1 m_2}{r_0} + \frac{Gm_1 m_2}{r_n} \ .$$

If one defines the gravitational potential energy V_k at t_k by

$$V_k = - \frac{Gm_1 m_2}{r_k} \ , \tag{2.18}$$

which is identical with the classical definition in continuous mechanics, then

$$W_n = -V_n + V_0 \ . \tag{2.19}$$

Hence, (2.11) and (2.19) imply, by the elimination of W_n, that

$$K_n + V_n = K_0 + V_0, \quad n = 1,2,3,\ldots \tag{2.20}$$

which is the law of conservation of energy.

Once again, it is worth noting that a telescopic sum, which in this case is

$$\sum_0^{n-1} [\frac{1}{r_k} - \frac{1}{r_{k+1}}] \ ,$$

had a key role in the discussion.

We wish to turn next to an actual computational example in which an orbit is constructed. However, as will be seen, we will have to be able to solve a nonlinear

algebraic system which is not trivial. For this reason, we summarize next a computer technique, called the generalized Newton's method [37] which will enable us
to solve our problem.

2.5   THE GENERALIZED NEWTON'S METHOD

Consider a system of k equations

$$f_1(x_1, x_2, x_3, \ldots, x_{k-1}, x_k) = 0$$

$$f_2(x_1, x_2, x_3, \ldots, x_{k-1}, x_k) = 0$$

$$f_3(x_1, x_2, x_3, \ldots, x_{k-1}, x_k) = 0 \tag{2.21}$$

$$\cdot$$
$$\cdot$$
$$\cdot$$

$$f_k(x_1, x_2, x_3, \ldots, x_{k-1}, x_k) = 0$$

in the k unknowns $x_1, x_2, x_3, \ldots, x_{k-1}, x_k$. A natural ordering of equations and a
structuring of the individual functions will be dictated by the physics of each
problem to be considered, as will be seen later. The particular generalization of
the classical one-dimensional Newton's method which will be applied to systems of
form (2.21) is called the generalized Newton's method and is described as follows.
Initially, guess an approximate solution $x_1^{(0)}, x_2^{(0)}, x_3^{(0)}, \ldots, x_{k-1}^{(0)}, x_k^{(0)}$. Again, as
will be indicated later, knowledge of the physics of a problem often enables one
to make this initial guess judiciously. Next, iterate with the recursion formulas

$$x_1^{(n+1)} = x_1^{(n)} - \omega \frac{f_1(x_1^{(n)}, x_2^{(n)}, x_3^{(n)}, \ldots, x_{k-1}^{(n)}, x_k^{(n)})}{\frac{\partial f_1}{\partial x_1}(x_1^{(n)}, x_2^{(n)}, x_3^{(n)}, \ldots, x_{k-1}^{(n)}, x_k^{(n)})} \tag{2.22}$$

$$x_2^{(n+1)} = x_2^{(n)} - \omega \frac{f_2(x_1^{(n+1)}, x_2^{(n)}, x_3^{(n)}, \ldots, x_{k-1}^{(n)}, x_k^{(n)})}{\frac{\partial f_2}{\partial x_2}(x_1^{(n+1)}, x_2^{(n)}, x_3^{(n)}, \ldots, x_{k-1}^{(n)}, x_k^{(n)})} \tag{2.23}$$

$$x_3^{(n+1)} = x_3^{(n)} - \omega \frac{f_3(x_1^{(n+1)}, x_2^{(n+1)}, x_3^{(n)}, \ldots, x_{k-1}^{(n)}, x_k^{(n)})}{\frac{\partial f_3}{\partial x_3}(x_1^{(n+1)}, x_2^{(n+1)}, x_3^{(n)}, \ldots, x_{k-1}^{(n)}, x_k^{(n)})} \tag{2.24}$$

$$\cdot$$
$$\cdot$$
$$\cdot$$

$$x_k^{(n+1)} = x_k^{(n)} - \omega \frac{f_k(x_1^{(n+1)}, x_2^{(n+1)}, x_3^{(n+1)}, \ldots, x_{k-1}^{(n+1)}, x_k^{(n)})}{\frac{\partial f_k}{\partial x_k}(x_1^{(n+1)}, x_2^{(n+1)}, x_3^{(n+1)}, \ldots, x_{k-1}^{(n+1)}, x_k^{(n)})} \tag{2.25}$$

where ω is a constant in the range

$$0 < \omega < 2. \tag{2.26}$$

Terminate the iteration when, for preassigned, positive ε,

$$|x_i^{(n+1)} - x_i^{(n)}| < \varepsilon, \quad i = 1,2,..,k. \tag{2.27}$$

Finally, check that $x_1^{(n+1)}, x_2^{(n+1)}, x_3^{(n+1)}, \ldots, x_k^{(n+1)}$ is a solution of (2.21) as an a posteriori verification that a solution of (2.21) exists.

Of particular importance in the above method are that new iterates are used in successive formulas immediately upon becoming available, and that no cumbersome matrix inversion routine is required, thus allowing k, the number of equations, to be unusually large.

2.6 AN ORBIT EXAMPLE

We turn now to an example of an orbit problem to illustrate the ideas developed thus far. In the notation of Section 2.4, let $Gm_1 = 1$, thus "normalizing" the units to be used. For initial conditions, let $x_0 = 0.50$, $y_0 = 0.00$, $v_{0,x} = 0.00$, $v_{0,y} = 1.63$. Had we not changed the units of measurement, then numbers would have to be given with more astronomical magnitudes. From (2.3)-(2.6) and (2.13)-(2.15), the equations of motion of the planet P can be written in the form

$$x_{k+1} - x_k - \frac{\Delta t}{2}(v_{k+1,x} + v_{k,x}) = 0 \tag{2.28}$$

$$y_{k+1} - y_k - \frac{\Delta t}{2}(v_{k+1,y} + v_{k,y}) = 0 \tag{2.29}$$

$$v_{k+1,x} - v_{k,x} + \frac{(x_{k+1}+x_k)\Delta t}{(x_k^2+y_k^2)^{\frac{1}{2}}(x_{k+1}^2+y_{k+1}^2)^{\frac{1}{2}}[(x_k^2+y_k^2)^{\frac{1}{2}}+(x_{k+1}^2+y_{k+1}^2)^{\frac{1}{2}}]} = 0 \tag{2.30}$$

$$v_{k+1,y} - v_{k,y} + \frac{(y_{k+1}+y_k)\Delta t}{(x_k^2+y_k^2)^{\frac{1}{2}}(x_{k+1}^2+y_{k+1}^2)^{\frac{1}{2}}[(x_k^2+y_k^2)^{\frac{1}{2}}+(x_{k+1}^2+y_{k+1}^2)^{\frac{1}{2}}]} = 0. \tag{2.31}$$

For $x_k, y_k, v_{k,x}$ and $v_{k,y}$ known, equations (2.28)-(2.31) are four equations in the four unknowns $x_{k+1}, y_{k+1}, v_{k+1,x}, v_{k+1,y}$. If these equations are denoted, respectively, by

$$f_1(x_{k+1}, y_{k+1}, v_{k+1,x}, v_{k+1,y}) = 0 \tag{2.32}$$

$$f_2(x_{k+1}, y_{k+1}, v_{k+1,x}, v_{k+1,y}) = 0 \tag{2.33}$$

$$f_3(x_{k+1}, y_{k+1}, v_{k+1,x}, v_{k+1,y}) = 0 \tag{2.34}$$

$$f_4(x_{k+1}, y_{k+1}, v_{k+1,x}, v_{k+1,y}) = 0, \tag{2.35}$$

where

$$f_1(x_{k+1}, y_{k+1}, v_{k+1,x}, v_{k+1,y}) = x_{k+1} - x_k - \frac{\Delta t}{2}(v_{k+1,x} + v_{k,x})$$

$$f_2(x_{k+1}, y_{k+1}, v_{k+1,x}, v_{k+1,y}) = y_{k+1} - y_k - \frac{\Delta t}{2}(v_{k+1,y} + v_{k,y})$$

$$f_3 = v_{k+1,x} - v_{k,x} + \frac{(x_{k+1} + x_k)\Delta t}{(x_k^2 + y_k^2)^{\frac{1}{2}}(x_{k+1}^2 + y_{k+1}^2)^{\frac{1}{2}}[(x_k^2 + y_k^2)^{\frac{1}{2}} + (x_{k+1}^2 + y_{k+1}^2)^{\frac{1}{2}}]}$$

$$f_4 = v_{k+1,y} - v_{k,y} + \frac{(y_{k+1} + y_k)\Delta t}{(x_k^2 + y_k^2)^{\frac{1}{2}}(x_{k+1}^2 + y_{k+1}^2)^{\frac{1}{2}}[(x_k^2 + y_k^2)^{\frac{1}{2}} + (x_{k+1}^2 + y_{k+1}^2)^{\frac{1}{2}}]} \quad ,$$

then, the Newtonian iteration formulas to be used for the solution of (2.28)-(2.31) are

$$x_{k+1}^{(n+1)} = x_{k+1}^{(n)} - \omega f_1(x_{k+1}^{(n)}, y_{k+1}^{(n)}, v_{k+1,x}^{(n)}, v_{k+1,y}^{(n)}) \tag{2.36}$$

$$y_{k+1}^{(n+1)} = y_{k+1}^{(n)} - \omega f_2(x_{k+1}^{(n+1)}, y_{k+1}^{(n)}, v_{k+1,x}^{(n)}, v_{k+1,y}^{(n)}) \tag{2.37}$$

$$v_{k+1,x}^{(n+1)} = v_{k+1,x}^{(n)} - \omega f_3(x_{k+1}^{(n+1)}, y_{k+1}^{(n+1)}, v_{k+1,x}^{(n)}, v_{k+1,y}^{(n)}) \tag{2.38}$$

$$v_{k+1,y}^{(n+1)} = v_{k+1,y}^{(n)} - \omega f_4(x_{k+1}^{(n+1)}, y_{k+1}^{(n+1)}, v_{k+1,x}^{(n+1)}, v_{k+1,y}^{(n)}) \quad , \tag{2.39}$$

since $\dfrac{\partial f_1}{\partial x_{k+1}} \equiv \dfrac{\partial f_2}{\partial y_{k+1}} \equiv \dfrac{\partial f_3}{\partial v_{k+1,x}} \equiv \dfrac{\partial f_4}{\partial v_{k+1,y}} \equiv 1$.   In iterating with (2.36)-(2.39), we begin, of course, with the given initial data x_0, y_0, $v_{0,x}$ and $v_{0,y}$. Then, each Newton iteration at time t_{k+1} is begun with the initial guess $x_{k+1}^{(0)} = x_k$, $y_{k+1}^{(0)} = y_k$, $v_{k+1,x}^{(0)} = v_{k,x}$, $v_{k+1,y}^{(0)} = v_{k,y}$, since these should be relatively close to x_{k+1}, y_{k+1}, $v_{k+1,x}$, $v_{k+1,y}$, respectively.

As a typical example of the calculations, planetary motion was generated with $\omega = 1$ and $\Delta t = 0.001$ up to $t_{350000} = 350$. The total computing time was under 5 minutes on the UNIVAC 1108. There were 86+ orbits, the 86th of which is shown in Figure 2.2. For this particular orbit, the period is $\tau = 4.05$ and the average of the absolute values of the x intercepts yields a semi-major axis length of 0.746.

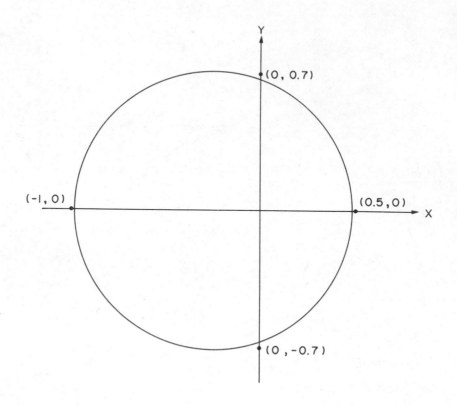

Fig. 2.2

2.7 GRAVITY REVISITED

Let us show now that our theory of gravitation is so general that, from it, we can deduce the results we already know about gravity. Of course, all we need do is prove that acceleration due to gravity is $a_k \equiv -32$ ft/sec^2, $k = 0,1,2,\ldots$. To do this, consider a circular earth of mass m_1, positioned at the origin of an XY coordinate system, as shown in Fig. 2.3. Let particle P be positioned with its center on the X-axis, as shown in Fig. 2.3, and let its mass m_2 be relatively small compared to that of m_1, so that the motion of the earth due to gravitational forces can be neglected. Then, since the motion of P is along the X-axis, we have, by (2.13),

$$F_{k,x} = -\frac{Gm_1 m_2 (x_{k+1}+x_k)}{r_k r_{k+1} (r_k+r_{k+1})} = -\frac{Gm_1 m_2 (x_{k+1}+x_k)}{x_k x_{k+1} (x_k+x_{k+1})} ,$$

which implies

$$F_{k,x} = - \frac{Gm_1 m_2}{x_k x_{k+1}} . \tag{2.40}$$

However, one also has

$$F_{k,x} = m_2 a_{k,x} , \tag{2.41}$$

so that, (2.40) and (2.41) imply

$$a_{k,x} = - \frac{Gm_1}{x_k x_{k+1}} . \tag{2.42}$$

Note that, from (2.42), the acceleration of P does not depend on the mass of P, but only on the mass of the earth.

Next, let us merely accept the astronomers' calculation of the mass of the earth [26], which, in grams, is approximately

$$m_1 \sim (0.598)10^{28} \; gr,$$

and also accept the usual estimate that the mean radius of the earth is 3959 miles. Recall also (2.2), that is, that in cgs units $G = (6.67)10^{-8}$. And for consistency and convenience, let us proceed now to the actual calculations using cgs units.

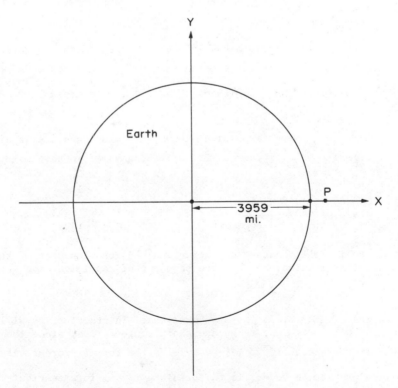

Fig. 2.3

From Figure 2.3 one realizes quickly that the size of the earth is so great that if, initially, P is close to the earth, then the distance P falls is <u>very small</u> compared to the radius of the earth. Thus, to a high degree of accuracy, $x_k \equiv 3959$ mi, k = 0,1,2,..., or, equivalently,

$$x_k = (6.373)10^8 \text{cm}, \quad k = 0,1,2,\ldots \; .$$

Substitution into (2.42) then yields

$$a_{k,x} \sim - \frac{(6.67)10^{-8}(0.598)10^{28}}{(6.373)^2 10^{16}} = - 982 \text{ cm/sec}^2. \tag{2.43}$$

Of course, changing cm into feet yields the equivalent result

$$a_{k,x} \sim -32 \text{ ft/sec}^2, \quad k = 0,1,2,\ldots \; . \tag{2.44}$$

2.8 CLASSICAL MOLECULAR FORCES

Physically, molecules behave in the following, simple way. Two molecules attract when relatively far apart and repel when relatively close, and repulsion is a much greater force than is attraction ([26], [50]). For example, the magnitude of the force between two molecules, each of unit mass, might have a formula like

$$F = - \frac{1}{r^6} + \frac{1}{r^{10}}, \tag{2.45}$$

where r is the distance between the molecules. From (2.45) it follows that $F = 0$ when $r = 1$, the attractive component $-\frac{1}{r^6}$ dominates when $r > 1$, and the repulsive component $\frac{1}{r^{10}}$ dominates when $r < 1$. In general, we proceed as follows.

Let particles P_1, P_2 have respective masses m_1, m_2. Then the local, molecular type force F which P_2 exerts on P_1 will be assumed to have magnitude F, given by

$$F = - \frac{Gm_1 m_2}{r^p} + \frac{Hm_1 m_2}{r^q}, \tag{2.46}$$

where G is called the constant of attraction, H the constant of repulsion, p the exponent of attraction, and q the exponent of repulsion, and where

$$q > p > 0, \quad G \geq 0, \quad H \geq 0. \tag{2.47}$$

The formulation (2.46)-(2.47) is so general that, in practice, we will be able to use it to deal with <u>both</u> long range and short range forces, since the particular choice $p = 2$, $G = (6.67)10^{-8}$, $H = 0$ yields the formula for gravitation.

Our immediate problem, however, is to formulate (2.46) in a fashion similar to (2.13)-(2.14) so that the resulting arithmetic equations conserve exactly the same energy as do their continuous counterparts. This formulation was developed recently and the proof of energy conservation is entirely analogous to that given for gravitation. The precise formulas are ([31]):

$$F_{k,x} = - \frac{Gm_1 m_2 \left[\sum_{j=0}^{p-2} (r_k^j r_{k+1}^{p-j-2})\right](x_{k+1}+x_k)}{r_k^{p-1} r_{k+1}^{p-1}(r_{k+1}+r_k)}$$

$$+ \frac{Hm_1 m_2 \left[\sum_{j=0}^{q-2} (r_k^j r_{k+1}^{q-j-2})\right](x_{k+1}+x_k)}{r_k^{q-1} r_{k+1}^{q-1}(r_{k+1}+r_k)} \qquad (2.48)$$

$$F_{k,y} = - \frac{Gm_1 m_2 \left[\sum_{j=0}^{p-2} (r_k^j r_{k+1}^{p-j-2})\right](y_{k+1}+y_k)}{r_k^{p-1} r_{k+1}^{p-1}(r_{k+1}+r_k)}$$

$$+ \frac{Hm_1 m_2 \left[\sum_{j=0}^{q-2} (r_k^j r_{k+1}^{q-j-2})\right](y_{k+1}+y_k)}{r_k^{q-1} r_{k+1}^{q-1}(r_{k+1}+r_k)} . \qquad (2.49)$$

2.9 REMARK

With regard to the motion of a single particle, it is worth noting that all the arithmetic, conservative formulas given thus far can, in fact, be derived from the following, single general formula [56]. Given any Newtonian potential $\phi(r)$, let

$$\vec{F}_k = - \frac{\phi(r_{k+1})-\phi(r_k)}{r_{k+1}-r_k} \cdot \frac{\vec{r}_{k+1}+\vec{r}_k}{r_{k+1}+r_k} \qquad (2.50)$$

Arithmetic formula (2.50) is conservative, independently of Δt, exactly as its continuous, limiting counterpart

$$\vec{F} = - \frac{\partial \phi}{\partial r} \frac{\vec{r}}{r} . \qquad (2.51)$$

Thus, the unifying concept which is the same in both arithmetic and continuous mechanics is the potential.

Chapter 3

The N-body Problem

3.1 INTRODUCTION

The discussion of Chapter 2 left a variety of interesting and important matters
unexplored. For example, we did not examine the possibility that several moving
bodies were interacting, or that there might have been other physical quantities,
besides energy, which were conserved. It is these matters which will be explored
in the present chapter. In order to simplify the mathematics of the discussion,
we will begin by a detailed study of the three-body problem, which is the prototype, nontrivial problem. The extension, then, to the interaction of any number of
bodies, under long and/or short range forces, will be straightforward.

3.2 THE THREE-BODY PROBLEM

Given the masses, position and velocity of three particles, the three-body problem
is to determine their motion if each is under the gravitational influence of the
other two. In analogy with the arithmetic formulation of Chapter II, we now proceed as follows.

For $\Delta t > 0$ and $t_k = k\Delta t$, $k = 0,1,2,\ldots$, and for each of $i = 1,2,3$, let
particle P_i of mass m_i be located at $\vec{r}_{i,k} = (x_{i,k}, y_{i,k})$, have velocity
$\vec{v}_{i,k} = (v_{i,k,x}, v_{i,k,y})$, and acceleration $\vec{a}_{i,k} = (a_{i,k,x}, a_{i,k,y})$ at time t_k. In
analogy with (2.3) and (2.4), let

$$\frac{\vec{v}_{i,k+1} + \vec{v}_{i,k}}{2} = \frac{\vec{r}_{i,k+1} - \vec{r}_{i,k}}{\Delta t} \ , \quad i = 1,2,3; \quad k = 0,1,2,\ldots, \tag{3.1}$$

$$\vec{a}_{i,k} = \frac{\vec{v}_{i,k+1} - \vec{v}_{i,k}}{\Delta t} \ , \quad i = 1,2,3; \quad k = 0,1,2,\ldots \ . \tag{3.2}$$

Of course, (3.1)-(3.2) differ from (2.3)-(2.4) only by the addition of the sub-
script i, which enables one to associate a given velocity and acceleration with
a particular particle in the system.

To relate force and acceleration, we assume a discrete Newtonian equation

$$\vec{F}_{i,k} = m_i \, \vec{a}_{i,k}; \quad i = 1,2,3; \quad k = 0,1,2,\dots, \tag{3.3}$$

where

$$\vec{F}_{i,k} = (F_{i,k,x}, F_{i,k,y}). \tag{3.4}$$

This time, the work W_n is defined by

$$W_n = \sum_{i=1}^{3} W_{i,n}, \tag{3.5}$$

where

$$W_{i,n} = \sum_{k=0}^{n-1} [(x_{i,k+1} - x_{i,k})F_{i,k,x} + (y_{i,k+1} - y_{i,k})F_{i,k,y}]. \tag{3.6}$$

The exact derivation which yielded (2.11) implies

$$W_{i,n} = \frac{m_i}{2}(v_{i,n,x}^2 + v_{i,n,y}^2) - \frac{m_i}{2}(v_{i,0,x}^2 + v_{i,0,y}^2), \tag{3.7}$$

so that if the kinetic energy $K_{i,k}$ of P_i at t_k is defined by

$$K_{i,k} = \frac{m_i}{2}(v_{i,k,x}^2 + v_{i,k,y}^2), \tag{3.8}$$

then

$$W_{i,n} = K_{i,n} - K_{i,0}. \tag{3.9}$$

Defining the kinetic energy K_k of the system at time t_k by

$$K_k = \sum_{i=1}^{3} K_{i,k} \tag{3.10}$$

yields, finally,

$$W_n = K_n - K_0. \tag{3.11}$$

Next, the precise structure of the force components of (3.4) is given as follows. If $r_{ij,k}$ is the distance between P_i and P_j at time t_k, then, in analogy with (2.13) and (2.14), set

$$F_{1,k,x} = -\frac{Gm_1 m_2 [(x_{1,k+1} + x_{1,k}) - (x_{2,k+1} + x_{2,k})]}{r_{12,k} r_{12,k+1}(r_{12,k} + r_{12,k+1})}$$

$$-\frac{Gm_1 m_3 [(x_{1,k+1} + x_{1,k}) - (x_{3,k+1} + x_{3,k})]}{r_{13,k} r_{13,k+1}(r_{13,k} + r_{13,k+1})} \tag{3.12}$$

$$F_{2,k,x} = - \frac{Gm_1 m_2 [(x_{2,k+1}+x_{2,k})-(x_{1,k+1}+x_{1,k})]}{r_{12,k} r_{12,k+1}(r_{12,k}+r_{12,k+1})}$$

$$- \frac{Gm_2 m_3 [(x_{2,k+1}+x_{2,k})-(x_{3,k+1}+x_{3,k})]}{r_{23,k} r_{23,k+1}(r_{23,k}+r_{23,k+1})} \tag{3.13}$$

$$F_{3,k,x} = - \frac{Gm_1 m_3 [(x_{3,k+1}+x_{3,k})-(x_{1,k+1}+x_{1,k})]}{r_{13,k} r_{13,k+1}(r_{13,k}+r_{13,k+1})}$$

$$- \frac{Gm_2 m_3 [(x_{3,k+1}+x_{3,k})-(x_{2,k+1}+x_{2,k})]}{r_{23,k+1} r_{23,k}(r_{23,k}+r_{23,k+1})} , \tag{3.14}$$

while $F_{1,k,y}$, $F_{2,k,y}$, $F_{3,k,y}$ are defined by interchanging y and x in (3.12), (3.13) and (3.14), respectively.

Each of (3.12)-(3.14) has two gravitational terms on the right-hand side because each particle is now influenced by two other particles.

3.3 CONSERVATION OF ENERGY

To establish the conservation of energy, consider again (3.5). Substitution of (3.12)-(3.14) and the corresponding formulas for $F_{1,k,y}$, $F_{2,k,y}$ and $F_{3,k,y}$ into (3.5) yields readily

$$W_n = - Gm_1 m_2 \sum_{k=0}^{n-1} \left(\frac{r_{12,k+1}-r_{12,k}}{r_{12,k} r_{12,k+1}} \right) - Gm_1 m_3 \sum_{k=0}^{n-1} \left(\frac{r_{13,k+1}-r_{13,k}}{r_{13,k} r_{13,k+1}} \right)$$

$$- Gm_2 m_3 \sum_{k=0}^{n-1} \left(\frac{r_{23,k+1}-r_{23,k}}{r_{23,k} r_{23,k+1}} \right)$$

$$= - Gm_1 m_2 \left(\frac{1}{r_{12,0}} - \frac{1}{r_{12,n}} \right) - Gm_1 m_3 \left(\frac{1}{r_{13,0}} - \frac{1}{r_{13,n}} \right)$$

$$- Gm_2 m_3 \left(\frac{1}{r_{23,0}} - \frac{1}{r_{23,n}} \right) .$$

Defining the potential energy $V_{ij,k}$ of the pair P_i and P_j at t_k by

$$V_{ij,k} = - G \frac{m_i m_j}{r_{ij,k}}$$

implies then that

$$W_n = V_{12,0} + V_{13,0} + V_{23,0} - V_{12,n} - V_{13,n} - V_{23,n} . \tag{3.15}$$

If the potential energy V_k of the system at time t_k is defined by

$$V_k = V_{12,k} + V_{13,k} + V_{23,k} \, ,$$

then (3.15) implies

$$W_n = V_0 - V_n \, . \tag{3.16}$$

Finally, elimination of W_n between (3.11) and (3.16) yields the desired result:

$$K_n + V_n = K_0 + V_0, \quad n = 1,2,3,\ldots \, .$$

3.4 SOLUTION OF THE DISCRETE THREE-BODY PROBLEM

For instructive purposes, let us consider now a particular three-body problem and show, in detail, how to solve it. The reasoning required for all other problems is analogous. Consider, therefore, as shown in Fig. 3.1, three particles P_1, P_2, P_3, of equal masses which are normalized so that

$$m_1 = m_2 = m_3 = 10, \quad G = 1. \tag{3.17}$$

Let the initial positions and velocities be given by $x_{1,0} = 0$, $y_{1,0} = 100$, $x_{2,0} = 100$, $y_{2,0} = 0$, $x_{3,0} = -100$, $y_{3,0} = 0$, $v_{1,0,x} = 0$, $v_{1,0,y} = -10$, $v_{2,0,x} = -10$, $v_{2,0,y} = 0$, $v_{3,0,x} = 9.9$, $v_{3,0,y} = 0$. From (3.1)-(3.4), one can, as in (2.28)-(2.31), rewrite the equations of motion as follows:

$$x_{i,k+1} - x_{i,k} - \frac{\Delta t}{2}(v_{i,k+1,x} + v_{i,k,x}) = 0, \quad i = 1,2,3 \tag{3.18}$$

$$y_{i,k+1} - y_{i,k} - \frac{\Delta t}{2}(v_{i,k+1,y} + v_{i,k,y}) = 0, \quad i = 1,2,3 \tag{3.19}$$

$$v_{1,k+1,x} - v_{1,k,x} + 10\Delta t \left\{ \frac{(x_{1,k+1} + x_{1,k}) - (x_{2,k+1} + x_{2,k})}{r_{12,k} r_{12,k+1}(r_{12,k} + r_{12,k+1})} \right.$$
$$\left. + \frac{(x_{1,k+1} + x_{1,k}) - (x_{3,k+1} + x_{3,k})}{r_{13,k} r_{13,k+1}(r_{13,k} + r_{13,k+1})} \right\} = 0 \tag{3.20}$$

$$v_{1,k+1,y} - v_{1,k,y} + 10\Delta t \left\{ \frac{(y_{1,k+1} + y_{1,k}) - (y_{2,k+1} + y_{2,k})}{r_{12,k} r_{12,k+1}(r_{12,k} + r_{12,k+1})} \right.$$
$$\left. + \frac{(y_{1,k+1} + y_{1,k}) - (y_{3,k+1} + y_{3,k})}{r_{13,k} r_{13,k+1}(r_{13,k} + r_{13,k+1})} \right\} = 0 \tag{3.21}$$

$$v_{2,k+1,x} - v_{2,k,x} + 10\Delta t \left\{ \frac{(x_{2,k+1} + x_{2,k}) - (x_{1,k+1} + x_{1,k})}{r_{12,k} r_{12,k+1}(r_{12,k} + r_{12,k+1})} \right.$$
$$\left. + \frac{(x_{2,k+1} + x_{2,k}) - (x_{3,k+1} + x_{3,k})}{r_{23,k} r_{23,k+1}(r_{23,k} + r_{23,k+1})} \right\} = 0 \tag{3.22}$$

$$v_{2,k+1,y} - v_{2,k,y} + 10\Delta t \left\{ \frac{(y_{2,k+1} + y_{2,k}) - (y_{1,k+1} + y_{1,k})}{r_{12,k} r_{12,k+1} (r_{13,k} + r_{13,k+1})} \right.$$

$$\left. + \frac{(y_{2,k+1} + y_{2,k}) - (y_{3,k+1} + y_{3,k})}{r_{23,k} r_{23,k+1} (r_{23,k} + r_{23,k+1})} \right\} = 0 \qquad (3.23)$$

$$v_{3,k+1,x} - v_{3,k,x} + 10\Delta t \left\{ \frac{(x_{3,k+1} + x_{3,k}) - (x_{1,k+1} + x_{1,k})}{r_{13,k} r_{13,k+1} (r_{13,k} + r_{13,k+1})} \right.$$

$$\left. + \frac{(x_{3,k+1} + x_{3,k}) - (x_{2,k+1} + x_{2,k})}{r_{23,k+1} r_{23,k} (r_{23,k} + r_{23,k+1})} \right\} = 0 \qquad (3.24)$$

$$v_{3,k+1,y} - v_{3,k,y} + 10\Delta t \left\{ \frac{(y_{3,k+1} + y_{3,k}) - (y_{1,k+1} + y_{1,k})}{r_{13,k} r_{13,k+1} (r_{13,k} + r_{13,k+1})} \right.$$

$$\left. + \frac{(y_{3,k+1} + y_{3,k}) - (y_{2,k+1} + y_{2,k})}{r_{23,k+1} r_{23,k} (r_{23,k} + r_{23,k+1})} \right\} = 0 \qquad (3.25)$$

where

$$r_{ij,k} = [(x_{i,k} - x_{j,k})^2 + (y_{i,k} - y_{j,k})^2]^{\frac{1}{2}} \qquad (3.26)$$

The solution of the twelve equations (3.18)-(3.25) for the twelve unknowns $x_{i,k+1}$, $y_{i,k+1}$, $v_{i,k+1,x}$, $v_{i,k+1,y}$, $i = 1,2,3$, for each value of $k = 0,1,2,$ ..., from the initial data is found by the generalized Newton's method with initial guess

$$x_{i,k+1}^{(0)} = x_{i,k}, \quad y_{i,k+1}^{(0)} = y_{i,k}, \quad v_{i,k+1,x}^{(0)} = v_{i,k,x}, \quad v_{i,k+1,y}^{(0)} = v_{i,k,y} \ .$$

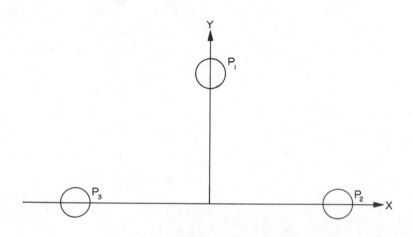

Fig. 3.1

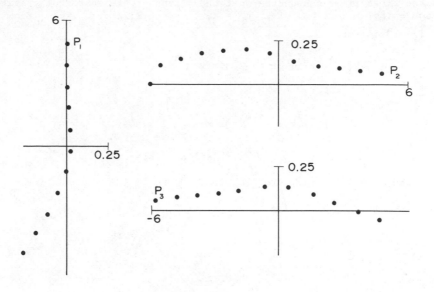

Fig. 3.2

In Fig. 3.2 are shown for $\Delta t = 0.1$ the deflections in the particles' trajectories from times t_{95} to t_{105}. The motion of each particle is shown separately and the labels P_i, $i = 1,2,3$ are affixed at their positions corresponding to t_{95}. The running time for one thousand time steps was under twenty seconds on the UNIVAC 1108.

3.5 CENTER OF GRAVITY

In the next three sections, we will study various interesting and important properties of three-body systems. These properties are not possessed by each individual particle, but are possessed by the system as a whole. We know, already, for example, that the total energy of a three-body system is conserved. But, no individual particle's energy need be conserved. A convenient place to begin is with the concept of the center of gravity.

At time t_k, let P_i of mass m_i be at $(x_{i,k}, y_{i,k})$, $i = 1,2,3$. Let

$$M = m_1 + m_2 + m_3$$

be the mass of the system. Then the unique point $(\bar{x}_k, \bar{y}_k)$ such that

$$M\bar{x}_k = m_1 x_{1,k} + m_2 x_{2,k} + m_3 x_{3,k}$$

$$M\bar{y}_k = m_1 y_{1,k} + m_2 y_{2,k} + m_3 y_{3,k}$$

(3.27)

is called the center of gravity, or the center of mass, of the system at time t_k.

Let us see first if we can deduce what the motion of the center of gravity of a three-body problem must be. Let $t_k = k\Delta t$, $k = 0, 1, 2, \ldots$. Then, from (3.3) and (3.12)-(3.14) one has

$$m_1 a_{1,k,x} + m_2 a_{2,k,x} + m_3 a_{3,k,x} = 0, \quad k \geq 0 \tag{3.28}$$

Hence,

$$m_1 (v_{1,k+1,x} - v_{1,k,x}) + m_2 (v_{2,k+1,x} - v_{2,k,x})$$
$$+ m_3 (v_{3,k+1,x} - v_{3,k,x}) = 0. \tag{3.29}$$

Summing both sides of (3.29) over k from 0 to $j - 1$, where $j \geq 1$, yields

$$m_1 (v_{1,j,x} - v_{1,0,x}) + m_2 (v_{2,j,x} - v_{2,0,x}) + m_3 (v_{3,j,x} - v_{3,0,x}) = 0. \tag{3.30}$$

However, since (3.30) is valid even if $j = 0$, it follows that

$$m_1 v_{1,j,x} + m_2 v_{2,j,x} + m_3 v_{3,j,x} = c_1, \quad j \geq 0 \tag{3.31}$$

where

$$c_1 = m_1 v_{1,0,x} + m_2 v_{2,0,x} + m_3 v_{3,0,x}. \tag{3.32}$$

Since (3.31) is valid for any j, it must be valid if j is replaced by $j + 1$, so that

$$m_1 v_{1,j+1,x} + m_2 v_{2,j+1,x} + m_3 v_{3,j+1,x} = c_1. \tag{3.33}$$

Addition of (3.31) and (3.33) then yields

$$m_1 \left[\frac{v_{1,j+1,x} + v_{1,j,x}}{2} \right] + m_2 \left[\frac{v_{2,j+1,x} + v_{2,j,x}}{2} \right]$$
$$+ m_3 \left[\frac{v_{3,j+1,x} + v_{3,j,x}}{2} \right] = c_1,$$

or, equivalently,

$$m_1 (x_{1,j+1} - x_{1,j}) + m_2 (x_{2,j+1} - x_{2,j})$$
$$+ m_3 (x_{3,j+1} - x_{3,j}) = c_1 \Delta t, \quad j \geq 0. \tag{3.34}$$

Summing both sides of (3.34) with respect to j from 0 to $n - 1$, for $n \geq 1$, yields

$$m_1 (x_{1,n} - x_{1,0}) + m_2 (x_{2,n} - x_{2,0}) + m_3 (x_{3,n} - x_{3,0}) = c_1 t_n. \tag{3.35}$$

However, (3.35) is valid also for $n = 0$, so that

$$m_1 x_{1,n} + m_2 x_{2,n} + m_3 x_{3,n} = c_1 t_n + c_2, \quad n \geq 0, \tag{3.36}$$

where

$$c_2 = m_1 x_{1,0} + m_2 x_{2,0} + m_3 x_{3,0}. \tag{3.37}$$

In a fashion analogous to the derivation of (3.36), it follows also that

$$m_1 y_{1,n} + m_2 y_{2,n} + m_3 y_{3,n} = d_1 t_n + d_2, \quad n \geq 0, \tag{3.38}$$

where

$$d_1 = m_1 v_{1,0,y} + m_2 v_{2,0,y} + m_3 v_{3,0,y}$$

$$d_2 = m_1 y_{1,0} + m_2 y_{2,0} + m_3 y_{3,0}.$$

Hence, (3.36) and (3.37) imply

$$M\bar{x}_n = c_1 t_n + c_2, \quad n \geq 0$$

$$M\bar{y}_n = d_1 t_n + d_2, \quad n \geq 0,$$

from which it follows that the motion of the center of gravity is along a straight line, with degeneracy occurring only in case $c_1^2 + d_1^2 \equiv 0$, in which case the center of gravity is always the point (c_2, d_2).

3.6 CONSERVATION OF LINEAR MOMENTUM

In studying the motion of an object, it is convenient to have a measure of how much force it takes to stop the object from its motion. The measure is called the object's linear momentum and is defined as follows. If particle P has mass m and velocity $\vec{v}$, then its linear momentum is defined as $m\vec{v}$. If, instead of one particle, there are more, say, three particles, then the linear momentum of the system is defined as the sum of the linear momenta of the three particles. If one now reexamines the discussion of the motion of the center of gravity in Section 3.6, one sees that one can deduce an interesting result from the equations developed there. Equation (3.31) and the corresponding equation for the y components are precise statements that, at any time, the linear momentum of the system is always the same as what it was at t_0. Thus, the motion of the system, however complex it may be, always conserves the sum of the linear momenta of its particles. This result is our second basic conservation law and is called the Law of Conservation of Linear Momentum.

The third, and final, conservation law, the Conservation of Angular Momentum, will be developed next.

3.7 CONSERVATION OF ANGULAR MOMENTUM

Suppose a bicycle wheel is set in a horizontal position and is made to rotate around its axle. Then a very interesting effect results, and it is most noticeable when the wheel is rotating very quickly. If one exerts a force perpendicular to the plane of the wheel, then the rotating wheel tends to resist this force. It seems to want to continue to rotate in its plane, and even though its motion is entirely two dimensional, it seems to push back on any force which is not in this plane of motion. Moreover, the greater its mass, or its radius, or its speed, the more marked is this tendency to resist. We say that this effect is due to its angular momentum. The same effect due to angular momentum is also noticeable when one ties a metal ball to the end of a string of length r and, by twirling, makes it rotate quickly in a plane, circular path of radius r. The ball resists any force perpendicular to its plane of motion. Indeed, the same effect is also apparent in planetary motion, since, for example, one can think of the above ball replaced by the earth and the string constraint replaced by the gravitational attraction to the sun.

In developing a formula for angular momentum, consider a particle P of mass m which, at t_k, is located at $\vec{r}_k$ and has velocity $\vec{v}_k$. From the above discussion, we know that, whatever else, the magnitude of the angular momentum of P should vary directly with its mass, its distance r to the center, and   its speed. It has been found that a very convenient way to do exactly this [69] is to define angular momentum as the vector $\vec{L}$ which is given by

$$\vec{L}_k = m(\vec{r}_k \times \vec{v}_k). \tag{3.39}$$

Let us show next that if one determines the angular momentum of each particle of a three-body problem, then the sum of these three angular momenta never changes. To do this, at time t_j let particle P_i of mass m_i be located at $\vec{r}_{ij}$, have velocity $\vec{v}_{i,j}$, and have angular momentum $\vec{L}_{i,j}$, that is,

$$\vec{L}_{i,j} = m_i(\vec{r}_{i,j} \times \vec{v}_{i,j}). \tag{3.40}$$

In the system of three particles, let the system angular momentum $\vec{L}_j$ at t_j be defined by

$$\vec{L}_j = \sum_{i=1}^{3} \vec{L}_{i,j} . \tag{3.41}$$

What we wish to show is that

$$\vec{L}_j = \vec{L}_0, \quad j = 1,2,3,\ldots, \tag{3.42}$$

and this is done as follows.

From (3.40) and the laws of vector cross products,

$$\vec{L}_{i,k+1} - \vec{L}_{i,k} = m_i(\vec{r}_{i,k+1} \times \vec{v}_{i,k+1}) - m_i(\vec{r}_{i,k} \times \vec{v}_{i,k})$$

$$= m_i[(\vec{r}_{i,k+1} - \vec{r}_{i,k}) \times (\frac{\vec{v}_{i,k+1} + \vec{v}_{i,k}}{2})$$

$$+ (\frac{\vec{r}_{i,k+1} + \vec{r}_{i,k}}{2}) \times (\vec{v}_{i,k+1} - \vec{v}_{i,k})].$$

From (3.1)-(3.4), then,

$$\vec{L}_{i,k+1} - \vec{L}_{i,k} = m_i [(\vec{r}_{i,k+1} - \vec{r}_{i,k}) \times (\frac{\vec{r}_{i,k+1} - \vec{r}_{i,k}}{\Delta t})$$

$$+ (\frac{\vec{r}_{i,k+1} + \vec{r}_{i,k}}{2}) \times (\vec{a}_{i,k} \Delta t)]$$

$$= \Delta t (\frac{\vec{r}_{i,k+1} + \vec{r}_{i,k}}{2}) \times \vec{F}_{i,k} .$$

For notational simplicity, set

$$\vec{T}_{i,k} = \frac{\vec{r}_{i,k+1} + \vec{r}_{i,k}}{2} \times \vec{F}_{i,k} , \tag{3.43}$$

so that

$$\vec{L}_{i,k+1} - \vec{L}_{i,k} = (\Delta t) \vec{T}_{i,k} . \tag{3.44}$$

Hence, if

$$\vec{T}_k = \sum_{i=1}^{3} \vec{T}_{i,k} , \tag{3.45}$$

then (3.41), (3.44), and (3.45) imply

$$\vec{L}_{k+1} - \vec{L}_k = (\Delta t) \vec{T}_k . \tag{3.46}$$

Now, if

$$\vec{T}_k = 0, \quad k = 0,1,2,\ldots \tag{3.47}$$

then

$$\vec{L}_{k+1} = \vec{L}_k, \quad k = 0,1,2,\ldots, \tag{3.48}$$

which implies (3.42), and the discussion would be complete. It remains for us to show then that, for the three-body problem, (3.47) is valid.

To do this observe that

$$\vec{T}_k = \vec{T}_{1,k} + \vec{T}_{2,k} + \vec{T}_{3,k}$$

$$= \frac{\vec{r}_{1,k+1} + \vec{r}_{1,k}}{2} \times \vec{F}_{1,k} + \frac{\vec{r}_{2,k+1} + \vec{r}_{2,k}}{2} \times F_{2,k} + \frac{\vec{r}_{3,k+1} + \vec{r}_{3,k}}{2} \times \vec{F}_{3,k} , \tag{3.49}$$

where $\vec{F}_{1,k}$, $\vec{F}_{2,k}$ and $\vec{F}_{3,k}$ are, by (3.12)-(3.14) and the corresponding formulas for the y-components:

$$\vec{F}_{1,k} = -\frac{Gm_1 m_2 (\vec{r}_{1,k+1} + \vec{r}_{1,k} - \vec{r}_{2,k+1} - \vec{r}_{2,k})}{r_{12,k} r_{12,k+1} (r_{12,k} + r_{12,k+1})}$$

$$- \frac{Gm_1 m_3 (\vec{r}_{1,k+1} + \vec{r}_{1,k} - \vec{r}_{3,k+1} - \vec{r}_{3,k})}{r_{13,k} r_{13,k+1} (r_{13,k} + r_{13,k+1})} \tag{3.50}$$

$$\vec{F}_{2,k} = -\frac{Gm_1 m_2 (\vec{r}_{2,k+1} + \vec{r}_{2,k} - \vec{r}_{1,k+1} - \vec{r}_{1,k})}{r_{13,k} r_{13,k+1} (r_{13,k} + r_{13,k+1})}$$

$$- \frac{Gm_2 m_3 (\vec{r}_{2,k+1} + \vec{r}_{2,k} - \vec{r}_{3,k+1} - \vec{r}_{3,k})}{r_{23,k+1} r_{23,k} (r_{23,k+1} + r_{23,k})} \tag{3.51}$$

$$\vec{F}_{3,k} = -\frac{Gm_1 m_3 (\vec{r}_{3,k+1} + \vec{r}_{3,k} - \vec{r}_{1,k+1} - \vec{r}_{1,k})}{r_{13,k+1} r_{13,k} (r_{13,k} + r_{13,k+1})}$$

$$- \frac{Gm_2 m_3 (\vec{r}_{3,k+1} + \vec{r}_{3,k} - \vec{r}_{2,k+1} - \vec{r}_{2,k})}{r_{23,k+1} r_{23,k} (r_{23,k+1} + r_{23,k})} \tag{3.52}$$

However, direct substitution of (3.50)-(3.52) into (3.49) yields, by the laws of vector cross products,

$$\vec{T}_k = \vec{0}, \quad k = 0,1,2,\ldots . \tag{3.53}$$

Thus (3.42), which is called the Law of Conservation of Angular Momentum, is valid.

3.8 THE N-BODY PROBLEM

All the conservation laws established for the three-body problem are <u>exactly</u> those of classical continuum mechanics. Moreover, in a fashion which is entirely analogous to that for the three-body problem, these laws can also be established for the interaction of N bodies in both the cases of long range and short range forces [31]. We will then not repeat these derivations, but will give a precise arithmetic formulation of the general N-body problem for later computational convenience. A FORTRAN program of this general formulation is given in Appendix 1.

Again, for positive time step Δt, let $t_k = k\Delta t$, $k = 0,1,2,\ldots$. At time t_k let particle P_i of mass m_i be located at $\vec{r}_{i,k} = (x_{i,k}, y_{i,k})$, have velocity $\vec{v}_{i,k} = (v_{i,k,x}, v_{i,k,y})$, and have acceleration $\vec{a}_{i,k} = (a_{i,k,x}, a_{i,k,y})$, for $i = 1,2,\ldots,N$. Position, velocity, and acceleration are assumed to be related by

$$\frac{\vec{v}_{i,k+1} + \vec{v}_{i,k}}{2} = \frac{\vec{r}_{i,k+1} - \vec{r}_{i,k}}{\Delta t} \tag{3.54}$$

$$\vec{a}_{i,k} = \frac{\vec{v}_{i,k+1} - \vec{v}_{i,k}}{\Delta t} \tag{3.55}$$

If $\vec{F}_{i,k} = (F_{i,k,x}, F_{i,k,y})$ is the force acting on P_i at time t_k, then force and acceleration are assumed to be related by the discrete dynamical equation

$$\vec{F}_{i,k} = m_i \vec{a}_{i,k} \tag{3.56}$$

In particular, we now choose $\vec{F}_{i,k}$ to have a component of attraction which behaves like $-G/r^p$ and a component of repulsion which behaves like H/r^q, where G, H, p, q, and r were described in Section 2.8 and are suitably constrained by (2.47). For this purpose, let $r_{i,j,k}$ be the distance between P_i and P_j at t_k. Then $\vec{F}_{i,k}$, the force exerted on P_i by the remaining particles, is defined, in analogy with (2.48) and (2.49), by

$$
\vec{F}_{i,k} = m_i \sum_{\substack{j=1 \\ j \neq i}}^{N} m_j \left\{ \left(- \frac{G \sum_{\xi=0}^{p-2} (r_{ij,k}^{\xi} r_{ij,k+1}^{p-\xi-2})}{r_{ij,k}^{p-1} r_{ij,k+1}^{p-1} (r_{ij,k+1} + r_{ij,k})} \right. \right.
$$
$$
\left. \left. + \frac{H \sum_{\xi=0}^{q-2} (r_{ij,k}^{\xi} r_{ij,k+1}^{q-\xi-2})}{r_{ij,k}^{q-1} r_{ij,k+1}^{q-1} (r_{ij,k+1} + r_{ij,k})} \right) (\vec{r}_{i,k+1} + \vec{r}_{i,k} - \vec{r}_{j,k+1} - \vec{r}_{j,k}) \right\},
$$
$$
i = 1,2,\ldots,N. \tag{3.57}
$$

3.9   REMARK

Only for simple forces, like gravity, do the continuous and the arithmetic approaches yield exactly the same dynamical behavior. In general ([57], [58]), the two approaches yield results which differ by terms of order $(\Delta t)^3$ in both position and velocity. Recently [58], new numerical formulas have been developed which increase this order of magnitude difference to any prescribed exponent, but, for these, the conservation of angular momentum of systems which have more than one particle is still to be proved.

Chapter 4

Conservative Models

4.1 INTRODUCTION

The arithmetic approach developed thus far lends itself naturally and consistently to discrete, or particle type, models of complex physical phenomena. These have been developed in both the conservative, implicit fashion and in the less expensive, nonconservative, explicit fashion. Viable discrete models have been developed for vibrating strings ([31], [85]); heat conduction and convection ([31], [34], [38], [42]); free surface, laminar and turbulent fluid flows ([36], [38], [41]-[45]); shock wave generation ([31], [76]); elastic vibration ([31], [34]); and evolution of planetary-type bodies [44]. The forces in the models are always consistent with the long and short range forces which actually prevail and the modeling applies with equal ease to both linear and nonlinear phenomena.

In this chapter we will concentrate on conservative models only. Whenever possible, the derived physical insights and advantages will be described. In Chapter 5, we will explore a nonconservative, but more highly economical approach, to discrete modeling, and will discuss the relative merits of the different approaches at that time.

4.2 THE SOLID STATE BUILDING BLOCK

Let us begin by developing a viable model of a solid. In doing so, we will attempt to simulate contemporary physical thought, in which molecules and atoms exhibit small vibrations in place. Moreover, since gravity will be of almost no significance in the phenomena to be discussed at first, we will consider now only local forces. Hence, consider first a system of only two particles, P_1 and P_2, of equal mass, which interact according to (3.57). Assume that the force between the particles is <u>zero</u>. Then, from (3.57),

$$\frac{-G \sum_{\xi=0}^{p-2} (r_{ij,k}^{\xi} \, r_{ij,k+1}^{p-\xi-2})}{r_{ij,k}^{p-1} r_{ij,k+1}^{p-1} (r_{ij,k+1} + r_{ij,k})} + \frac{H \sum_{\xi=0}^{q-2} (r_{ij,k}^{\xi} \, r_{ij,k+1}^{q-\xi-2})}{r_{ij,k}^{q-1} r_{ij,k+1}^{q-1} (r_{ij,k+1} + r_{ij,k})} = 0 \qquad (4.1)$$

32

But, if there is zero force between the two particles, then $r_{ij,k} = r_{ij,k+1}$, so set $r_{ij,k} = r_{ij,k+1} = r$ in (4.1) to yield

$$\frac{-G \sum_{\xi=0}^{p-2} r^{p-2}}{r^{2p-2}} + \frac{H \sum_{\xi=0}^{q-2} r^{q-2}}{r^{2q-2}} = 0. \tag{4.2}$$

Since, by (2.47), $q > p$, then

$$- Gr^{-p}(p-1) + Hr^{-q}(q-1) = 0, \tag{4.3}$$

or, finally,

$$r = \left[\frac{H(q-1)}{G(p-1)}\right]^{1/(q-p)}. \tag{4.4}$$

Consider next a system of only three particles, P_1, P_2 and P_3, of equal masses, and assume that no force acts between any two of the particles. Then the distance between any two of the particles is given, again, by (4.4). Such a configuration is therefore exceptionally stable and will be called a triangular building block.

When considering a solid we will decompose it into triangular building blocks. Then, by an appropriate choice of parameters, the force on any particle of a triangular block due to more distant particles will be made small, thus achieving the small vibrations desired. To illustrate, let the six particles P_1, P_2, P_3, P_4, P_5, P_6 be located at the vertices of the four triangular building blocks of the triangular region OAB, shown in Fig. 4.1. Assume that $m_i \equiv 1$, $G = H = 1$, $p = 7$, and $q = 10$, so that $r = \sqrt[3]{1.5}$. The particles' initial positions are, then,

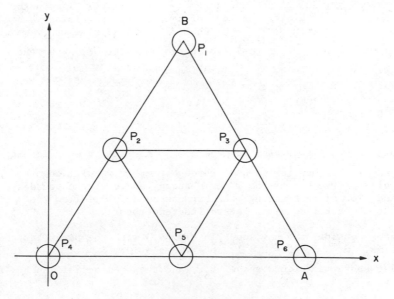

Fig. 4.1

$$(x_1, y_1) = (1.14471, 1.98270), \quad (x_2, y_2) = (0.57236, 0.99135)$$

$$(x_3, y_3) = (1.71707, 0.99135), \quad (x_4, y_4) = (0,0)$$

$$(x_5, y_5) = (1.14471, 0), \qquad (x_6, y_6) = (2.28943, 0).$$

Assign to each particle a $\vec{0}$ initial velocity. Finally, let particles P_4 and P_6 be fixed by defining the total force on each of these particles to be $\vec{0}$ and allow the remaining particles to move under force law (3.57). For $\Delta t = 0.05$ and for 2500 time steps, the motions of P_1 and P_5 exhibited small oscillations in the vertical direction only, while P_2 and P_3 exhibited small two dimensional oscillations. The maximum distance, for example, that P_1 moved from its initial position was approximately 0.02, and this occurred at approximately every one hundred time steps. The running time on the UNIVAC 1108 was 4 minutes and the program used was that in Appendix 1.

4.3   FLOW OF HEAT IN A BAR

Let us now develop the basic concepts of discrete conductive heat transfer by concentrating on the prototype problem of heat flow in a bar. Physically, the problem is formulated as follows. Let the region bounded by rectangle OABC, as shown in Fig. 4.2, represent a bar. Let $|OA| = a$, $|OC| = c$. A section of the boundary of the bar is heated. The problem is to describe the flow of heat through the bar.

Our discrete approach to the problem proceeds as follows. First, subdivide the given region into triangular building blocks, one such possible subdivision of which is shown in Fig. 4.3 for the parameter choices $m_i \equiv 1$, $G = H = 1$, $p = 7$, $q = 10$, $a \sim 11$, $c \sim 2$. In this case, from (4.4), $r \sim 1.1447142426$.

Now, by <u>heating</u> a section of the boundary of the bar, we will mean <u>increasing the velocity</u>, and hence the kinetic energy, of some of the particles whose centers are on OABC. By the <u>temperature</u> $T_{i,k}$ of particle P_i at time t_k, we will mean the following. Let M be a fixed positive integer and let $K_{i,k}$ be the kinetic energy of P_i at t_k. Then $T_{i,k}$ is defined by

$$T_{i,k} = \frac{1}{M} \sum_{j=k-M+1}^{k} K_{i,j}, \qquad (4.5)$$

which is, of course, the arithmetic mean of P_i's kinetic energies at M consecutive time steps. By the <u>flow</u> of heat through the bar we will mean the transfer to other particles of the bar of the kinetic energy added at the boundary. Finally, to follow the flow of heat through the bar one need only follow the motion of each particle and, at each time step, record its temperature.

To illustrate, consider the bar shown in Fig. 4.3 with the parameter choices given above, that is, $m_i \equiv 1$, $G = H = 1$, $p = 7$, $q = 10$, $a \sim 11$, $c \sim 2$. Assume that a strong heat source is placed above P_6, and then removed, in such a fashion that

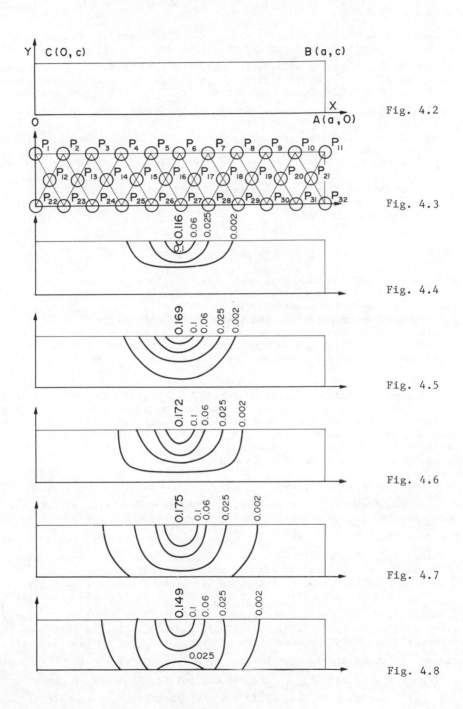

Fig. 4.2

Fig. 4.3

Fig. 4.4

Fig. 4.5

Fig. 4.6

Fig. 4.7

Fig. 4.8

$$\vec{v}_{5,0} = (\frac{-\sqrt{2}}{2}, \frac{-\sqrt{2}}{2}), \quad \vec{v}_{6,0} = (0,-1), \quad \vec{v}_{7,0} = (\frac{\sqrt{2}}{2}, \frac{-\sqrt{2}}{2}),$$

while all other inital velocities are $\vec{0}$. With regard to temperature calculation, assume that the velocities of all particles prior to t_0 were $\vec{0}$. As regards the choice of M, which is relatively arbitrary, set $M = 20$. From the resulting calculations with $\Delta t = 0.025$, Figs. 4.4-4.8 show the constant temperature contours $T = 0.1, 0.06, 0.025, 0.002$ at t_5, t_{10}, t_{15}, t_{20} and t_{25}, respectively. The resulting wave motion is clear and Fig. 4.8 exhibits wave reflection.

Other heat transfer concepts can be defined now in the same spirit as above, as follows. A side of a bar is insulated means that the bar particles cannot transfer energy across this side of the bar to particles outside the bar, as in the above example, while melting is the result of adding so much heat that various particles attain sufficient speeds which break the bonding effect of (3.57). Such problems are of wide interest but will require the introduction of gravity, which will be done in the next chapter.

4.4   OSCILLATION OF AN ELASTIC BAR

Next, let us develop the basic mechanisms of discrete elasticity by concentrating on the vibration of an elastic bar. The problem is formulated physically as follows. Let the region bounded by rectangle OABC, as shown in Fig. 4.2, represent a bar which can be deformed, and which, after deformation, tends to return to its original shape. The problem is to describe the motion of such a bar after an external force, which has deformed the bar, is removed. Equivalently, the problem is to describe the motion of an elastic bar after release from a position of tension.

Our discrete approach proceeds as follows. The given region is first subdivided into triangular building blocks. Then, deformation results in the compression of certain particles and the stretching apart of others. Release from a position of deformation, or tension, results, by (3.57), in repulsion between each pair of particles which has been compressed and attraction between each pair which has been stretched, the net effect being the motion of the bar.

As a particular example, let $m_i \equiv 1$, $p = 7$, $q = 10$, $G = 425$, $H = 1000$, and $\Delta t = .025$. From (4.7), $r = 1.52254$. Consider, for variety, the thirty particle bar which results by deleting P_{11} and P_{32} from the configuration of Fig. 4.3. The particles P_1, P_{12}, and P_{22}, whose respective coordinates are $(0, 2.63711)$, $(.76127, 1.31855)$, and $(0,0)$, are to be held fixed throughout. This is done by defining the total force on each of these two particles to be $\vec{0}$. In order to obtain an initial position of tension like that shown in Fig. 4.9(a), first set P_{13}, P_{14}, P_{15}, P_{16}, P_{17}, P_{18}, P_{19}, P_{20} and P_{21} at $(2.28357, 1.29198)$, $(3.80538, 1.26541)$, $(5.32632, 1.18573)$, $(6.84052, 1.02658)$, $(8.33992, .76219)$, $(9.81058, .36813)$, $(11.23199, -.17750)$, $(12.57631, -.89228)$, and $(13.80807, -1.78721)$, respectively. Any two consecutive points P_k, P_{k+1}, $k = 13,14,\ldots,20$, are positioned r units apart. The points $P_2 - P_{10}$ and $P_{23} - P_{31}$ are positioned as follows: P_{k-10} and P_{k+11} are the two points which are r units from both P_k and P_{k+1} for each of $k = 12,13,\ldots,20$. Each consecutive pair of points in the $P_2 - P_{10}$ sets are then separated by a distance greater than r, while each

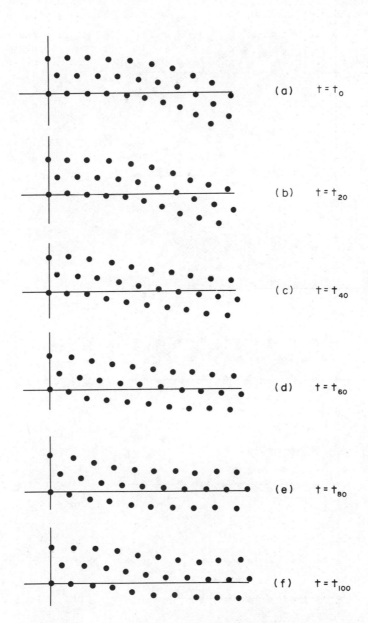

(a) $t = t_0$

(b) $t = t_{20}$

(c) $t = t_{40}$

(d) $t = t_{60}$

(e) $t = t_{80}$

(f) $t = t_{100}$

Fig. 4.9

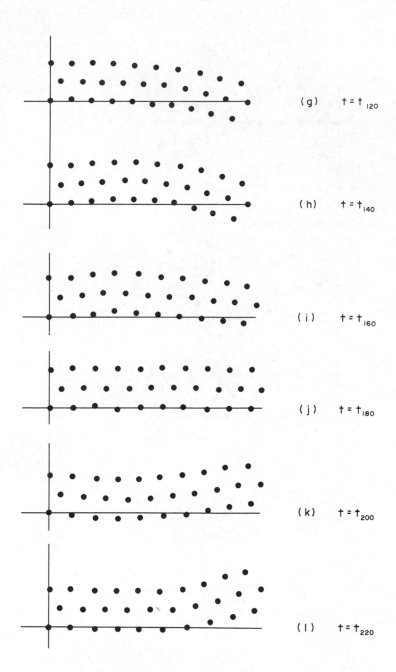

(g) $t = t_{120}$

(h) $t = t_{140}$

(i) $t = t_{160}$

(j) $t = t_{180}$

(k) $t = t_{200}$

(l) $t = t_{220}$

Fig. 4.9 completed

consecutive pair of points in the P_{23} - P_{31} set are separated by a distance less
than r. Thus, the points P_2 - P_{10} are in a stretched position, while the points
P_{23} - P_{31} are compressed.

From the initial position of tension shown in Fig. 4.9(a), the oscillatory motion of
the bar is determined from (3.54)-(3.57) with all initial velocities set at $\vec{0}$.
The upward swing of the bar was plotted automatically at every twenty time steps
and is shown in Fig. 4.9(a-1) from t_0 to t_{220}. It is of interest to note that as
the bar moves, each row of particles exhibits wave oscillation and reflection.
Engineers have been aware of such waves on the surfaces of vibrating materials for
some time.

4.5 LAMINAR AND TURBULENT FLUID FLOWS

As a final example of a discrete conservative model, we examine the flow of a
liquid out of a nozzle. By restricting attention, at present, to the immediate
flow, only, out of the nozzle, we can continue to neglect the effect of gravity.

Consider, then, a two dimensional liquid in motion, a small portion of which is
shown in Fig. 4.10. Let particles P_1 - P_{11} be called the first row, P_{12} - P_{23}
the second row, and P_{24} - P_{34} the third row. In (3.57), let $G \equiv H \equiv m_i \equiv 1$,
$i = 1,2,\ldots,34$, and $p = 7$, $q = 10$. The initial positions of P_1 - P_{34} are set
so that P_{13} - P_{22} are centers of regular hexagons of radii $r = (1.5)^{1/3}$, while
the remaining particles are centered at the vertices of the hexagons. This choice
of r is, of course, by (4.4), one of relative configuration stability.

The motion of the particles will be completely determined once we fix the initial
velocities $v_{i,0,x}$ and $v_{i,0,y}$. To do this, let us suppose that the particles
have just been emitted horizontally from a nozzle. If this were the case, then
$v_{i,0,x}$ would dominate $v_{i,0,y}$. Moreover, not all particles would have exactly
the same velocities because of possible collisions with the nozzle housing, and so
forth. So, let us choose

$$v_{i,0,x} = V + \varepsilon_{i,1}, \quad v_{i,0,y} = \varepsilon_{i,2}, \quad i = 1,2,\ldots,34 \qquad (4.6)$$

where V is a parameter which assures relatively horizontal motion, while $\varepsilon_{i,1}$
and $\varepsilon_{i,2}$ are relatively small random numbers which give the particles small per-
turbations from purely horizontal motion. For simplicity, let the computer generate all the $\varepsilon_{i,1}$ and $\varepsilon_{i,2}$ in a random fashion so that

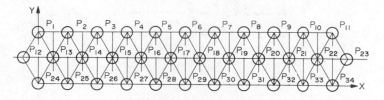

Fig. 4.10

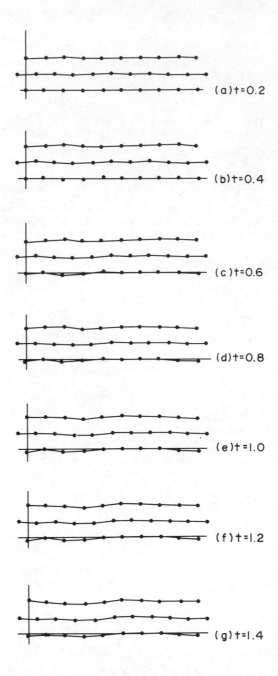

(a) t = 0.2

(b) t = 0.4

(c) t = 0.6

(d) t = 0.8

(e) t = 1.0

(f) t = 1.2

(g) t = 1.4

Fig. 4.11

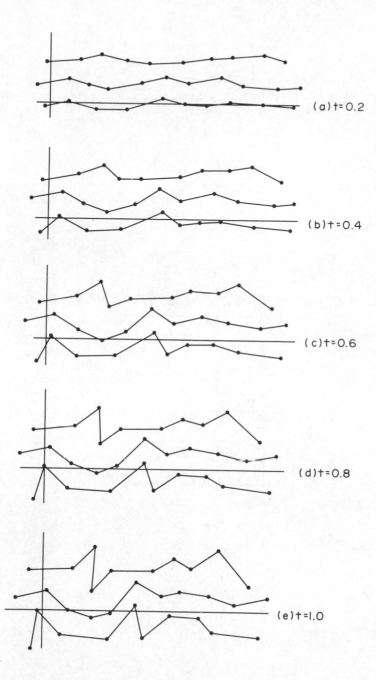

(a) t=0.2

(b) t=0.4

(c) t=0.6

(d) t=0.8

(e) t=1.0

Fig. 4.12

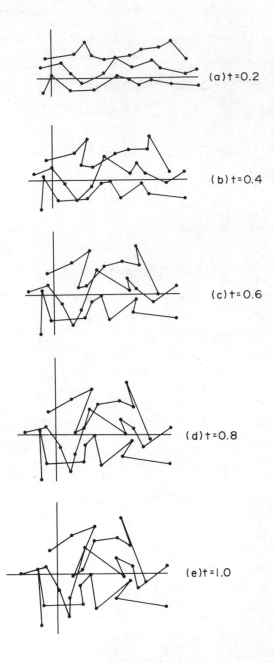

(a) t = 0.2

(b) t = 0.4

(c) t = 0.6

(d) t = 0.8

(e) t = 1.0

Fig. 4.13

$$|\varepsilon_{i,j}| \leq (1\%)V = \frac{V}{100}, \quad j = 1,2. \tag{4.7}$$

In the examples which follow, our interest will center on increasing values of V
and on initial time steps only. The random numbers are generated by the computer
only once and independently of V. However, in each example, these numbers are
rescaled proportionately so that the maximum $|\varepsilon_{i,j}|$ always gives <u>equality</u> in
(4.7).

Figure 4.11 shows the particle motion for V = 50 and Δt = 0.02 at t = 0.2,
0.4, 0.6, 0.8, 1.0, 1.2 and 1.4. A gentle wave motion develops in each row while
the rows maintain their relative positions. A flow of this nature is said to be
laminar. Figure 4.12 shows the motion for V = 300 and Δt = 0.02 at t = 0.2,
0.4, 0.6, 0.8 and 1.0. Repulsion between the particles has assumed a greater significance and, though the rows still maintain their relative positions, the motion
is becoming more <u>chaotic</u>. Figure 4.13 shows the motion for V = 1000 with
Δt = 0.01 at t = 0.2, 0.4, 0.6, 0.8 and 1.0. So much motion results that the
choice Δt = 0.1 was necessary for the convergence of the generalized Newton's
method. Here, the laminar character of the flow has disappeared in that the rows
no longer maintain their relative positions, and the motion becomes extremely
chaotic, or, more descriptively, turbulent. Thus, with the increase in velocity,
particles can come nearer to other particles, which results in increased repulsive
forces and more complex motion.

Intuitively, turbulence is often thought of as a type of fluid flow which is characterized by the rapid appearance and disappearance of many small vortices, or
whirlpools. If one defines a vortex as a counterclockwise, or a clockwise, motion
exhibited by three or more relatively close particles, as shown in Fig. 4.14, then,
indeed, such configurations do appear in our example for V = 1000, do break down
quickly due to the very large effects of repulsion, and then do reappear in different particle groupings.

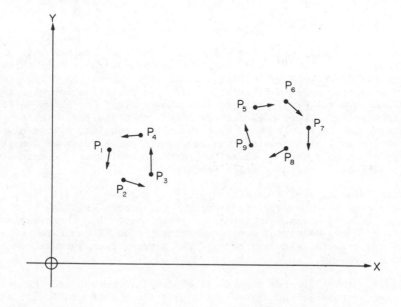

Fig. 4.14

Chapter 5

Nonconservative Models

5.1 INTRODUCTION

In each of the N-body, conservative models of Chapter 4, the implicit structure of
the equations required that a system of nonlinear, algebraic equations be solved at
each time step. Such a requirement is relatively expensive, even for moderate
values of N. One can, however, still formulate and study discrete models by using
explicit formulas, which will be illustrated by several examples in the present
chapter. Economy will then be obtained at the expense of exact conservation.
Often, however, we will be able to develop a reasonable nonconservative model by
concentrating on that portion of a natural phenomenon which is, itself, nonconservative. In Sections 5.4-5.7 we will demonstrate how to incorporate, in addition,
long range forces into the models.

5.2 SHOCK WAVES

In contrast with a liquid, a gas has relatively few particles per unit of volume.
Consider, then, a gas as shown in a long tube in Fig. 5.1(a). Into this tube in-
sert a piston, as shown in Fig. 5.1(b). If one first moves the piston down the
tube slowly, then, as shown in Fig. 5.1(c), the gas particles increase in density
per unit volume in a relatively uniform way. However, if, as shown in Fig. 5.1(d),
the piston is moved at a very high rate of speed, then gas particles compact on the
cylinder head, with the result that the original gas consists of two distinct portions, one with a very high density, the other with about the same density as at
the start. The boundary between these two portions, which is shown as a dotted
line in Fig. 5.1(d), is called a shock wave.

More generally, a shock wave can be thought of as follows. Assign to a given gas a
positive measure of average particle density. Let a body B pass through the gas
at a very high rate of speed. In certain regions about B, there may occur sets
of gas particles whose densities are not average. Then, a boundary between sets of
particles with average density and those with "greater than average" density is
called a shock wave.

Let us illustrate this "greater than average" density concept and the development
of a shock wave by considering next a particular shock tube problem. Consider the
tube configuration in Fig. 5.1(b). For convenience, a coordinate system will be

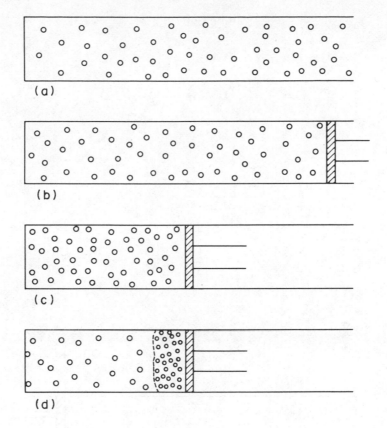

Fig. 5.1

fixed relative to the piston head, as shown in Fig. 5.2, so that the particles will
be considered to be in motion relative to the piston. Let the tube be 100 units
long, so that $AO = 100$, and 10 units high, so that $AB = 10$. Now, every
$\Delta t = 0.01$ seconds, let a column of particles, each of radius $r = 0.35$ and of <u>unit</u>
mass m, enter the tube at AB. Each such column is determined as follows. At
each time t_k, each position $(-100, n+\frac{1}{2})$, $n = 0,1,2,...,9$, is either filled by
a particle or left vacant by a random process, like the toss of a coin. Once it
has been determined that a particle P_i is at such an initial location, then its
initial velocity $\vec{v}_{i,0}$ is determined by

$$\vec{v}_{i,0} = (100+\varepsilon_{i,1}, \varepsilon_{i,2}),$$

where $\varepsilon_{i,1}$ and $\varepsilon_{i,2}$ are selected at random, but are small in magnitude relative
to 100, thus assuring that the gas has a relative high speed in a relatively uni-
form direction. Because of the high speeds and short time durations in shock wave
development, gravity will be neglected in the formulation. Allowing a repulsive
force to simulate the effects of particle collision, we now assume an N-body formulation in which the position $(x_{i,k}, y_{i,k})$ and velocity $(v_{i,k,x}, v_{i,k,y})$ of P_i
at time t_k are generated explicitly by (3.54) and (3.55) in the particular form

$$v_{i,k+1,x} = v_{i,k,x} + a_{i,k,x}\Delta t \tag{5.1}$$

$$v_{i,k+1,y} = v_{i,k,y} + a_{i,k,y}\Delta t \tag{5.2}$$

$$x_{i,k+1} = x_{i,k} + \frac{\Delta t}{2}(v_{i,k+1,x}+v_{i,k,x}) \tag{5.3}$$

$$y_{i,k+1} = y_{i,k} + \frac{\Delta t}{2}(v_{i,k+1,y}+v_{i,k,y}) \tag{5.4}$$

and by

$$a_{i,k,x} = \frac{1}{m}\sum_{\substack{j=1\\j\neq i}}^{N}\frac{(x_{i,k}-x_{j,k})H}{(r_{ij,k}+\xi)^{q+1}} \tag{5.5}$$

$$a_{i,k,y} = \frac{1}{m}\sum_{\substack{j=1\\j\neq i}}^{N}\frac{(y_{i,k}-y_{j,k})H}{(r_{ij,k}+\xi)^{q+1}} \tag{5.6}$$

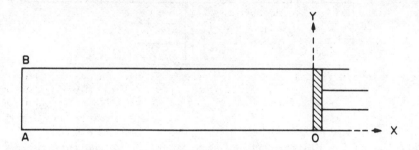

Fig. 5.2

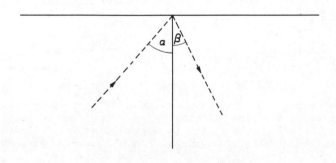

Fig. 5.3

where

$$r_{ij,k} = \text{distance between } P_i \text{ and } P_j, \text{ in the tube, at } t_k \qquad (5.7)$$

$$N = \text{total number of particles in the tube at } t_k, \text{ and} \qquad (5.8)$$

$$H = \begin{cases} 0, & \text{if } r_{ij,k} \geq 2r = 0.7, \\ \\ 1, & \text{if } r_{ij,k} < 2r = 0.7. \end{cases} \qquad (5.9)$$

The parameter ξ is a measure of how close the centers of two particles can come and serves to conserve mass.

If and when a particle impacts on either the top or the bottom of the tube, or on the piston head, we will assume that it rebounds, as shown in Fig. 5.3 with

$$\beta = \alpha \pm \gamma. \qquad (5.10)$$

The quantity γ is determined at random in the range $0 \leq \gamma \leq \frac{\pi}{40}$, subject to the restriction $0 \leq \beta \leq \frac{\pi}{2}$. If this last restriction is satisfied for both choices of sign in (5.10), then the sign is to be determined at random. If the incident speed is $|v_i|$, while the reflected speed is $|v_r|$, it will be assumed that

$$|v_r| = 0.2 |v_i|,$$

which can be interpreted as a transference of kinetic energy from the particles of the gas to the particles of the container. Note that the damping factor, which is 0.2 in this case, depends entirely on the nature of the boundary surface.

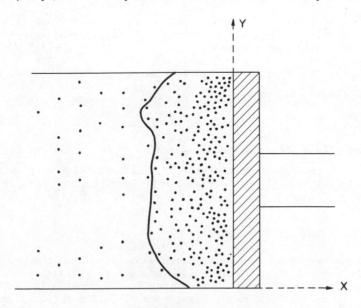

Fig. 5.4

For the above simple formulation with $q = 1$ and $\xi = 0.1$, Fig. 5.4 shows the shock wave structure at time t_{160} , that is, after 1.6 seconds, when "greater than average" density is defined to mean that the distance between a particle and at least five other particles is less than unity.

It should be noted that the heating of the walls of the tube, a phenomenon of fundamental importance in shock tube generation, is usually too difficult to incorporate into continuous models.

5.3 THE LEAP-FROG FORMULAS

Extensive calculation with (5.1)-(5.4) reveals that they may require severe stability restrictions [19]. In anticipation, then, of developing complex models which incorporate both long and short range forces, we will wish to avoid stability problems, and these can, in general, be so avoided by the use of a higher order set of formulas called the leap-frop formulas [31]. These are given as follows and will be used throughout the remainder of the chapter.

For positive time step Δt , let $t_k = k\Delta t$, $k = 0,1,\ldots$. For $i = 1,2,\ldots,N$, let particle P_i have mass m_i and at time t_k let P_i be located at $\vec{r}_{i,k}$ $= (x_{i,k}, y_{i,k})$, have velocity $\vec{v}_{i,k} = (v_{i,k,x}, v_{i,k,y})$ and have acceleration $\vec{a}_{i,k}$ $= (a_{i,k,x}, a_{i,k,y})$. Then the leap-frog formulas, which relate position, velocity, and acceleration, are as follows:

$$\vec{v}_{i,\frac{1}{2}} = \vec{v}_{i,0} + \frac{\Delta t}{2}\vec{a}_{i,0} \tag{5.11}$$

$$\vec{v}_{i,k+\frac{1}{2}} = \vec{v}_{i,k-\frac{1}{2}} + (\Delta t)\vec{a}_{i,k}, \quad k = 1,2,\ldots \tag{5.12}$$

$$\vec{r}_{i,k+1} = \vec{r}_{i,k} + (\Delta t)\vec{v}_{i,k+\frac{1}{2}}, \quad k = 0,1,2,\ldots \tag{5.13}$$

The name "leap-frog" is derived by the way the position and velocity are defined at alternate, sequential time points.

If $\vec{F}_{i,k}$ is the force acting on P_i at time t_k , where $\vec{F}_{i,k} = (F_{i,k,x}, F_{i,k,y})$, then we will assume, as usual, that

$$\vec{F}_{i,k} = m_i\vec{a}_{i,k} \tag{5.14}$$

Of course, once an exact structure is given to $\vec{F}_{i,k}$, then the motion of each particle will be determined recursively by (5.11)-(5.14) from prescribed initial data. The special structures to be used are described in detail in the sections which follow.

5.4 THE STEFAN PROBLEM

Stefan problems are free boundary problems associated with the processes of melting of solids and crystallizing of liquids. The name is derived from one of the early studies of Arctic ice formation. Specifically, the problem is that of describing the changing shape of the boundary between the liquid portion and solid portion when either, or both, of the above processes are occurring.

The Stefan problem is of such difficulty that classical mathematical studies have been restricted largely to existence, uniqueness, and asymptotic behavior of a one dimensional model based on the linear heat equation ([28], [55], [61]). The recent availability of high speed digital computation has led to the development of finite difference, variational, finite element, and Chebyshev-series numerical methods for approximating the solution of the problem ([10], [18], [21], [23], [51], [73]). Most of these methods have been applied to one dimensional problems only, and, in all cases, only to the linear heat equation. Indirect efforts, which avoid computing the free boundary, itself, have also been developed and require either the numerical solution of a nonlinear partial differential equation ([14], [65]) or the numerical solution of a variational inequality [2].

In this section we will develop a direct, particle approach to Stefan problems. Though attention is restricted to the melting phenomenon only, it is equally applicable to crystallization.

Using the leap-frog formulation of Section 5.2, we assume first that the local force $(F_{i,k,x}, F_{i,k,y})$ exerted on P_i by P_j is given by

$$\overline{F}_{i,k,x} = \left[\frac{-Gm_i m_j}{r_{ij,k}^p} + \frac{Hm_i m_j}{r_{ij,k}^q} \right] \frac{x_{i,k} - x_{j,k}}{r_{ij,k}} \tag{5.15}$$

$$\overline{F}_{i,k,y} = \left[\frac{-Gm_i m_j}{r_{ij,k}^p} + \frac{Hm_i m_j}{r_{ij,k}^q} \right] \frac{y_{i,k} - y_{j,k}}{r_{ij,k}}. \tag{5.16}$$

The total force $(F_{i,k,x}^*, F_{i,k,y}^*)$ on P_i due to $N - 1$ other particles is given by

$$F_{i,k,x}^* = \sum_{\substack{j=1 \\ j \neq i}}^{N} \overline{F}_{i,k,x}; \quad F_{i,k,y}^* = \sum_{\substack{j=1 \\ j \neq i}}^{N} \overline{F}_{i,k,y}. \tag{5.17}$$

Finally, since gravity is essential in the process of melting, we include it as follows:

$$F_{i,k,x} = F_{i,k,x}^*; \quad F_{i,k,y} = F_{i,k,y}^* - 32m_i. \tag{5.18}$$

Thus, (5.14) is now determined completely.

We cannot, however, use the simplistic approach to the construction of a solid which was demonstrated in Chapter 4. The reason is that when we melt a solid, gravity must be present if the fluid part is to flow downward. For this reason, we must now show how to generate a solid which has short range forces and gravity. We will then do this first by reconsidering the discussion of Section 4.3.

As shown in Fig. 5.5, let P_1, P_2, P_3 be located at the vertices of an isosceles triangle in which $|P_1 P_2| = |P_1 P_3| = r$. In order to develop a strong, three particle bond, we will neglect all time dependences at present so that, for example, r, d, and h, as shown in Fig. 5.5, are constant and satisfy

$$h^2 + d^2 = r^2.$$

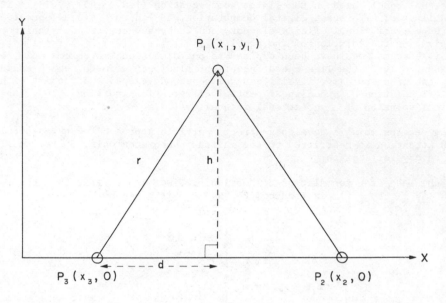

Fig. 5.5

Now, if P_1, P_2 and P_3 were strongly bonded, then all forces on these particles would be close to zero in magnitude. These forces, from (5.18), are, in general, given by

$$F_x(P_1) = \frac{-Gm_1m_2}{r_{12}{}^p}\frac{x_1-x_2}{r_{12}} + \frac{Hm_1m_2}{r_{12}{}^q}\frac{x_1-x_2}{r_{12}}$$

$$- \frac{Gm_1m_3}{r_{13}{}^p}\frac{x_1-x_3}{r_{13}} + \frac{Hm_1m_3}{r_{13}{}^q}\frac{x_1-x_3}{r_{13}} \qquad (5.19)$$

$$F_x(P_2) = \frac{-Gm_1m_2}{r_{12}{}^p}\frac{x_2-x_1}{r_{12}} + \frac{Hm_1m_2}{r_{12}{}^q}\frac{x_2-x_1}{r_{12}}$$

$$- \frac{Gm_2m_3}{r_{23}{}^p}\frac{x_2-x_3}{r_{23}} + \frac{Hm_2m_3}{r_{23}{}^q}\frac{x_2-x_3}{r_{23}} \qquad (5.20)$$

$$F_x(P_3) = \frac{-Gm_1m_3}{r_{13}^p} \frac{x_3-x_1}{r_{13}} + \frac{Hm_1m_3}{r_{13}^q} \frac{x_3-x_1}{r_{13}}$$

$$- \frac{Gm_2m_3}{r_{23}^p} \frac{x_3-x_2}{r_{23}} + \frac{Hm_2m_3}{r_{23}^q} \frac{x_3-x_2}{r_{23}} \tag{5.21}$$

$$F_y(P_1) = \frac{-Gm_1m_2}{r_{12}^p} \frac{y_1-y_2}{r_{12}} + \frac{Hm_1m_2}{r_{12}^q} \frac{y_1-y_2}{r_{12}}$$

$$- \frac{Gm_1m_3}{r_{13}^p} \frac{y_1-y_3}{r_{13}} + \frac{Hm_1m_3}{r_{13}^q} \frac{y_1-y_3}{r_{13}} - 32m_1 \tag{5.22}$$

$$F_y(P_2) = \frac{-Gm_1m_2}{r_{12}^p} \frac{y_2-y_1}{r_{12}} + \frac{Hm_1m_2}{r_{12}^q} \frac{y_2-y_1}{r_{12}}$$

$$- \frac{Gm_2m_3}{r_{23}^p} \frac{y_2-y_3}{r_{23}} + \frac{Hm_2m_3}{r_{23}^q} \frac{y_2-y_3}{r_{23}} - 32m_2 \tag{5.23}$$

$$F_y(P_3) = \frac{-Gm_1m_3}{r_{13}^p} \frac{y_3-y_1}{r_{13}} + \frac{Hm_1m_3}{r_{13}^q} \frac{y_3-y_1}{r_{13}}$$

$$- \frac{Gm_2m_3}{r_{23}^p} \frac{y_3-y_2}{r_{23}} + \frac{Hm_2m_3}{r_{23}^q} \frac{y_3-y_2}{r_{23}} - 32m_3 . \tag{5.24}$$

Next, for clarity let us consider the particular parameter choices

$$p = 4, \quad q = 6, \quad d = 0.6, \quad h = 0.8, \quad r = 1.0, \quad m_1 = m_2 = m_3 = m, \tag{5.25}$$

the reasoning being completely analogous for other sets of choices. Then (5.19)–(5.24) reduce to

$$F_x(P_1) = 0 \tag{5.26}$$

$$F_x(P_2) = -Gm^2(.6) + Hm^2(.6) - \frac{Gm^2}{(1.2)^4} + \frac{Hm^2}{(1.2)^6} \tag{5.27}$$

$$F_x(P_3) = -F_x(P_2) \tag{5.28}$$

$$F_y(P_1) = 2[-Gm^2(.8) + Hm^2(.8)] - 32m \tag{5.29}$$

$$F_y(P_2) = -Gm^2(-.8) + Hm^2(-.8) - 32m \tag{5.30}$$

$$F_y(P_3) = F_y(P_2).$$ (5.31)

If, as shown in Fig. 5.5, we now assume that P_2 and P_3 are fully supported below, which will be implemented later by strongly damped reflection, then the forces $F_y(P_2)$ and $F_y(P_3)$ in (5.30) and (5.31) will be of no consequence. Thus, we concentrate only on parameter choices which result in each of (5.26)-(5.29) being zero. Hence, we need only consider $F_x(P_2) = 0$ and $F_y(P_1) = 0$, or equivalently,

$$-[(1.2)^2 + (.6)(1.2)^6]G + [1 + (.6)(1.2)^6]H = 0$$

$$- G \qquad\qquad\qquad + H = \frac{20}{m},$$

the solution of which, in terms of m, is given by

$$G = \frac{1395.7952}{11m}, \quad H = G + \frac{20}{m}.$$ (5.32)

Thus, as a particular example, for unit mass and for G and H rounded to integers which are divisible by five, one has from (5.32)

$$m = 1, \quad G = 125, \quad H = 145.$$ (5.33)

From (5.32) and (5.33), one can construct additional approximations like

$$m = \frac{1}{5}, \quad G = 625, \quad H = 725$$ (5.34)

$$m = \frac{1}{20}, \quad G = 2500, \quad H = 2900.$$ (5.35)

In order to show next how to construct solids, we will restrict attention to two-dimensional configurations only and will always call these bodies. Triangular building blocks will be used in the construction of bodies. Of course, all the ideas to be developed extend in a natural way dimensionwise.

For simplicity and clarity, in this section we will consider only a triangular shaped body, so that, beginning with $N = 28$, let the particles $P_1 - P_{28}$ be arranged as shown in Fig. 5.6. The triangular building blocks each have base length 1.2 and equal arm length 1.0. The initial positions are given precisely in Table 5.1 and all initial velocities are taken to be zero.

Let us fix the parameters by (5.25) and (5.35) so that each building block whose base is on the X axis should be a bonded unit. It is not to be implied, however, that the entire body is in equilibrium, nor that the remaining building blocks are bonded units, since (5.35) followed from a variety of restrictions and assumptions. Nevertheless, each of the upper building blocks is supported by blocks beneath it, and the bonding constants H and G are sufficiently large so that the effect of gravity due to the total mass of the body should not be too great. Thus, if our intuition is correct, the body should be close to an equilibrium state. Our method of achieving such a desired state, will be to let the particles interact according to (5.11)-(5.18) with $\Delta t = 10^{-4}$ and to converge to equilibrium by themselves. There is, of course, no hope of finding an analytical solution of this N-body problem.

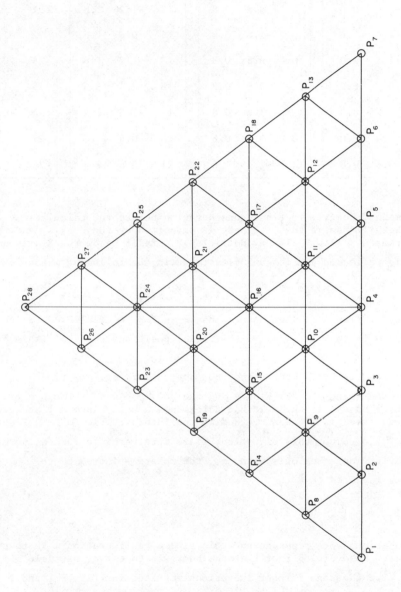

Fig. 5.6

TABLE 5.1   Initial Positions of Particles P_i : Triangular Body

i	$x_{i,0}$	$y_{i,0}$	i	$x_{i,0}$	$y_{i,0}$	i	$x_{i,0}$	$y_{i,0}$	i	$x_{i,0}$	$y_{i,0}$
1	-3.6	0.0	8	-3.0	0.8	15	-1.2	1.6	22	1.8	2.4
2	-2.4	0.0	9	-1.8	0.8	16	0.0	1.6	23	-1.2	3.2
3	-1.2	0.0	10	-0.6	0.8	17	1.2	1.6	24	0.0	3.2
4	0.0	0.0	11	0.6	0.8	18	2.4	1.6	25	1.2	3.2
5	1.2	0.0	12	1.8	0.8	19	-1.8	2.4	26	-0.6	4.0
6	2.4	0.0	13	3.0	0.8	20	-0.6	2.4	27	0.6	4.0
7	3.6	0.0	14	-2.4	1.6	21	0.6	2.4	28	0.0	4.8

There remains, however, a final consideration before the calculations can be exe-
cuted. We will assume that the body is supported by the X-axis, and this will be
implemented as follows. If any particle P_i falls below the X-axis at time t_k,
then it is reflected and its velocity reset in the following manner.

$$x_{i,k} \to x_{i,k}; \quad y_{i,k} \to -y_{i,k}; \quad v_{i,k,x} \to 0, \quad v_{i,k,y} \to (.1)v_{i,k,y}.$$

With this modification, then, particles $P_1 - P_{28}$ were allowed to interact in
accordance with (5.11)-(5.18), with initial positions those of Table 5.1, and with
initial velocities all zero.

The first results were interesting, but not satisfying. By time t_{10000} all the
particles had fallen to the X-axis and the body had gone through a transformation
like that of honey, in which it had deformed gradually into a flat surface. The
problem was that gravity was overcoming the interparticle forces of attraction and
repulsion. To remedy this situation, since interparticle forces increase with m^2
while gravity increases only with m, the mass was increased to $\frac{1}{4}$, so that
(5.35) was replaced by

$$m = \frac{1}{4}, \quad G = 2500, \quad H = 2900.$$

The calculations were repeated and this time a configuration like that in Fig. 5.6
did result. However, to test that the body was cohesive, particle P_{19} was
removed. Two thousand further iterations resulted in P_{23}, P_{26} and P_{28} sliding
down to replace P_{19}, P_{23} and P_{26}, respectively. Thus, the body simulated a pile
of sand. To remedy this situation the mass was increased to unity, so that (5.35)
was replaced by

$$m = 1, \quad G = 2500, \quad H = 2900 \qquad\qquad (5.36)$$

and the calculations were repeated again. The results now simulated a gravitational collapse in which each particle showed exceptionally strong attraction to
all other particles. Indeed, the entire body initially rose upward and gradually
settled down into a triangular body similar to, but smaller than, that shown in
Fig. 5.6. To test that the resulting body was cohesive, particle P_{19} was removed
again. Two thousand further iterations showed that P_1, P_8 and P_{14} had moved up
the side of the triangle while P_{23}, P_{26}, and P_{28} had moved down, thereby filling
in the void left by P_{19}. Thus, the resulting body was cohesive, but not in the
way one would expect a large solid body to be. The problem was that P_1 and P_{28},
for example, were attracting each other very strongly, whereas, for a large body,
this would not happen, since interparticle forces are local.

There are then two possible ways to remedy this last problem. One can take N
much larger than 28. In this way the distance between the furthest separated particles increases and the resulting interparticle force becomes negligible. To
economize, however, the following alternative was used with N = 28. We merely set
$\overline{F}_{i,k,x}$ and $\overline{F}_{i,k,y}$ in (5.15) and (5.16) equal to zero if $|P_iP_j| > 1.5$, thus
keeping these forces local. With this modification and with (5.36) the calcula-
tions were repeated again and were finally successful. To describe the results.
we let $K(t_k) = K_k$ be the system's kinetic energy at t_k. K_k will be used as a
measure of the body's relative state of equilibrium, since an absolute state is not
attainable directly on a computer.

By time t_{10000}, the system kinetic energy had decreased, in an oscillatory fashion, which is indicated by the sequence: $K_{500} = 436$, $K_{1000} = 487$, $K_{1500} = 211$,
$K_{2000} = 345$, $K_{2500} = 234$, $K_{3000} = 371$, $K_{3500} = 302$, $K_{4000} = 237$, $K_{4500} = 153$,
$K_{5000} = 195$, $K_{5500} = 195$, $K_{6000} = 230$, $K_{6500} = 122$, $K_{7000} = 91$, $K_{7500} = 94$,
$K_{8000} = 78$, $K_{8500} = 108$, $K_{9000} = 80$, $K_{9500} = 51$, $K_{100000} = 74$. The resulting
body is shown in Fig. 5.7 and all positions and velocities are recorded in Table
5.2. Particle oscillations showed changes in position coordinates in only the
third decimal place. Removal of particle P_{19} resulted in no additional struc-
tural movement in two thousand additional time steps.

Let us consider now the effect of placing a heat source near, but to the right of
the uppermost particle P_{28}, in Fig. 5.6. Suppose the heat source is at
(0.2, 5.7). Heat will be transferred to the body by increasing the velocity, and
hence the kinetic energy, of various body particles as follows. At each time step
t_k, consider the distance R_i of particle P_i to the heat source. If $R_i \geq 2.5$,
then P_i's velocity is left unchanged. If $R < 2.5$, then its velocity components
are reset to

$$\vec{v}_{i,k} = \left(v_{i,k,x} + \frac{\Gamma}{R_i^2} \frac{(x_{i,k}-0.2)}{R_i}, v_{i,k,y} + \frac{\Gamma}{R_i^2} \frac{(y_{i,k}-5.7)}{R_i} \right), \qquad (5.37)$$

so that the intensity varies as an inverse square, with Γ being a positive vari-
ation constant, and along the line joining the center of the particle and the center of the heat source, itself. In order to simulate a gradual increase of heat at the
source, itself, Γ is increased with time in the following fashion. Particle
motions are calculated with $\Gamma = 0.008$ for 50000 time steps, then with

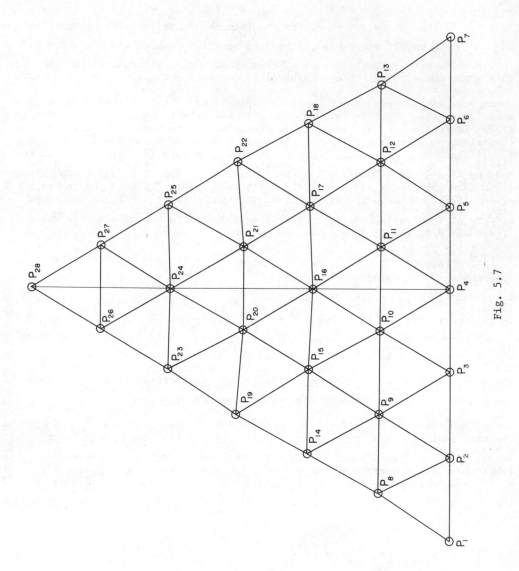

Fig. 5.7

TABLE 5.2 Equilibrium Positions and Velocities of Particles P_i : Triangular Body

i	x	y	v_x	v_y
1	-3.2652	0.0000	0.0000	0.0005
2	-2.1924	0.0000	0.0000	0.0065
3	-1.0895	0.0000	0.0000	0.0018
4	0.0000	0.0000	0.0000	0.0016
5	1.0895	0.0000	0.0000	0.0018
6	2.1924	0.0000	0.0000	0.0065
7	3.2652	0.0000	0.0000	0.0005
8	-2.6754	0.8909	-0.9437	1.0055
9	-1.6310	0.8990	2.0010	3.7973
10	-0.5445	0.9032	4.1844	-0.9128
11	0.5445	0.9032	-4.1844	-0.9128
12	1.6310	0.8990	-2.0010	3.7973
13	2.6754	0.8909	0.9437	1.0055
14	-2.1366	1.8374	0.9267	0.4191
15	-1.0763	1.8146	-2.5412	-2.1506
16	0.0000	1.7923	0.0000	-1.5731
17	1.0763	1.8146	2.5412	-2.1506
18	2.1366	1.8374	-0.9267	0.4191
19	-1.6674	2.7530	-0.9468	0.6165
20	-0.5441	2.6879	-1.8075	2.1719
21	0.5441	2.6879	1.8075	2.1719
22	1.6674	2.7530	0.9468	0.6165
23	-1.0877	3.6478	0.2061	-2.6777
24	0.0000	3.6174	0.0000	0.4901
25	1.0877	3.6478	-0.2061	-2.6777
26	-0.5536	4.5068	-0.5620	0.7410
27	0.5536	4.5068	0.5620	0.7410
28	0.0000	5.4351	0.0000	-2.8452

Γ = .0095 for 30000 additional time steps, and finally with Γ = 0.015 for 20000 time steps. The total physical time of the 100000 time steps is 10 seconds.

The question of whether or not a particle is a fluid particle or a solid particle will be determined, at present, in a simplistic way which relies on observation, from the computations, of the breaking of a bonded triple of particles. (A more refined approach will be described in the next section.) Note first that for the triangular solid of Fig. 5.7, the distance between any two particles in a triangular bond of three particles never exceeds 1.5, since $\vec{F}^* \equiv \vec{0}$ if $r_{ij,k} > 1.5$. In addition, if one defines the <u>average</u> linear velocity $v^*_{i,k}$ of P_i at t_k by

$$v^*_{i,k} = \frac{\vec{r}_{i,k} - \vec{r}_{i,k-1000}}{0.1}, \tag{5.38}$$

which is a convenient measure of the <u>gross</u>, rather than the local, motion of the particle, then $v^*_{i,k} > 7.5$ invariably yielded the breaking of a three-particle bond. Thus, we define a particle P_i at t_k to be a solid particle if both the following are valid:

(a) P_i and two other particles are vertices of a triangle for which each side has length less than 1.5,

(b) $|v^*_{j,k}| \le 7.5$ for each particle of the triangle in (a).

Otherwise, P_i is said to be a fluid particle.

Using the above criteria, Fig. 5.8 shows the melting of the triangular solid from $t_{80000} - t_{100000}$. The boundary between the solid and liquid portion is delineated by drawing line segments between all solid particles which are separated by a distance less that 1.5 units. Fluid particles with average speeds greater than 7.5 units have directional arrows attached, while those with velocities less than 7.5 do not.

Figure 5.8(a) shows the slight deformation of the solid at t_{80000} due to the accumulation of heat, while (b) shows a more rapid and more intensive deformation by the time t_{82000}. Melting appears in Fig. 5.8(c) at t_{84000} and accumulates to a flow down the left side by the time t_{86000}, as shown in (d). Figure 5.8(e), at t_{87000}, shows that P_1 has moved to the left sufficiently to enable P_{19} to bond with P_1 and P_8. However, the direction of P_{28} suggests that this bond cannot endure. In addition, as the particles in the upper right hand section of the solid begin to regroup, P_{25} has converted its potential energy into kinetic energy and has become a fluid particle. At this time, t_{87000}, no particle is receiving any heat from the source because all particles are now at least 2.25 units from the source. In (f), at t_{88000}, one sees three interesting formations. First, the bond which P_{19} has formed with P_1 and P_8 is beginning to break. Second, the bottom row has buckled, as was beginning to be evident in Fig. 5.8(e), a phenomenon which results from the limitation of horizontal motion along the base. Third, particles P_{17}, P_{22}, P_{21}, P_{27} and P_{16}, though each a solid particle, are

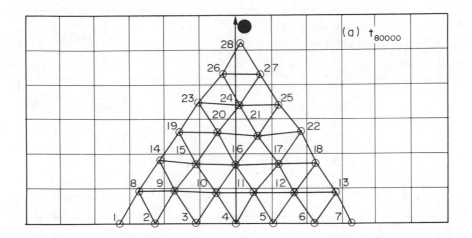

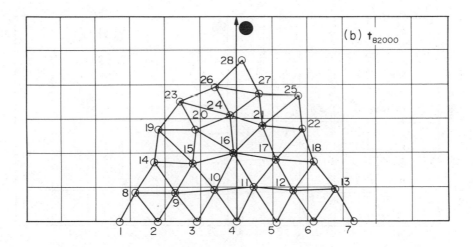

Fig. 5.8

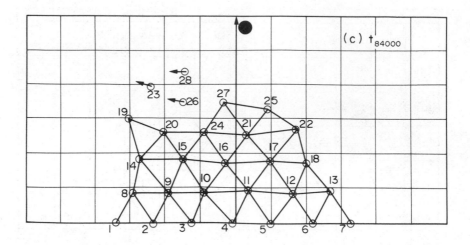

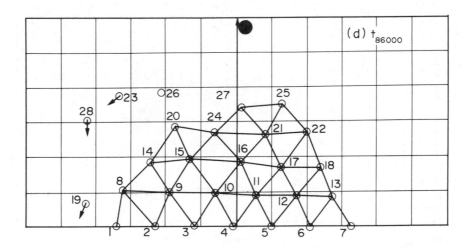

Fig. 5.8 – continued

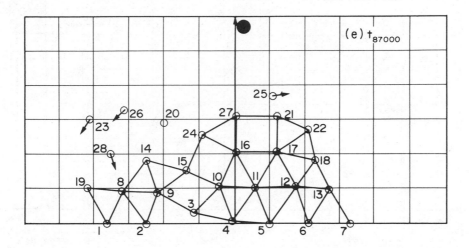

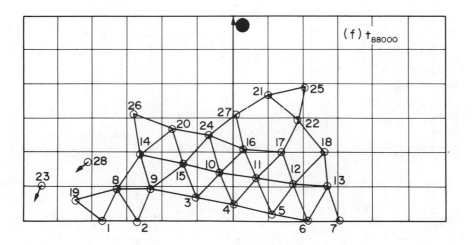

Fig. 5.8 - continued

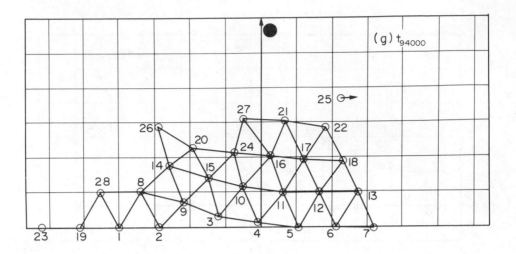

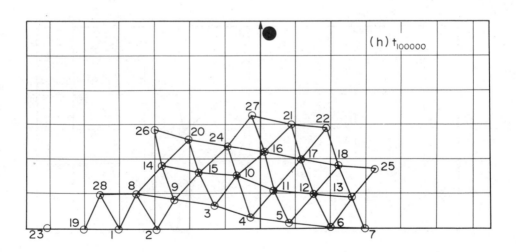

Fig. 5.8 - completed

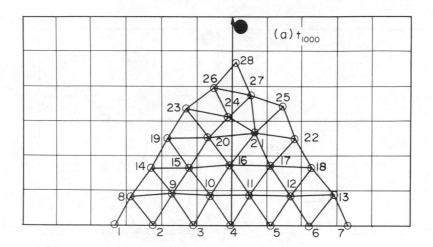

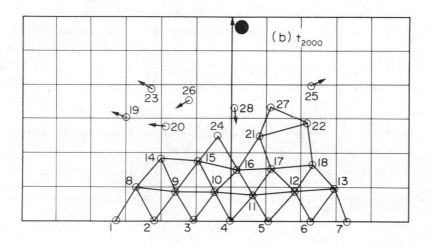

Fig. 5.9

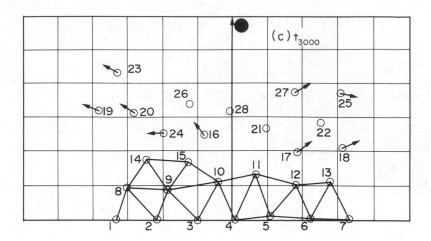

5.9 - completed

also particles which surround a relative void. Mutual attraction must then lead to a collapse into the void and the formation of additional bonds. In Fig. 5.8(g) at t_{94000} we see that P_{19} has formed a new bond with P_1 and P_{28}, while P_{17}, P_{22}, P_{21}, P_{27} and P_{16} have formed additional bonds, as predicted, but thereby leaving P_{25} unbonded, and hence a fluid particle. P_{25} then falls under the motion of gravity until it rebonds with P_{13} and P_{18}, as shown in Fig. 5.8(h) at t_{100000}, which is the final equilibrium formation. Only P_{23} remains as a fluid particle.

The melting and resolidifying processes in this example resemble the melting of a wax-like solid.

A more intensive type of heating is shown in Fig. 5.9. In this example, the heat source at (0.2, 5.8) is applied with $\Gamma = 0.02$ immediately and a splashing, or splattering, effect results quickly. Fig. 5.9(c) is especially revealing of this motion.

5.5 EVOLUTION OF PLANETARY TYPE BODIES

In Section 5.4 we considered a phenomenon in which the long range force was gravity. In this section, we will consider a phenomenon in which the long range force is gravitation. We will do this by modeling the evolution of a planetary type body from a swirling, molten mass.

Using the leap frog formulas of Section 5.2, we assume now that the force $(\overline{F}_{i,k,x}, \overline{F}_{i,k,y})$ exerted on P_i by P_j is given by

$$\overline{F}_{i,k,x} = \left[-\frac{G}{r_{ij,k}^{p}} + \frac{H}{r_{ij,k}^{q}} - \frac{G*}{r_{ij,k}^{2}} \right] \frac{m_i m_j (x_{i,k} - x_{j,k})}{r_{ij,k}} \qquad (5.39)$$

$$\overline{F}_{i,k,y} = \left[-\frac{G}{r_{ij,k}^{p}} + \frac{H}{r_{ij,k}^{q}} - \frac{G*}{r_{ij,k}^{2}} \right] \frac{m_i m_j (y_{i,k} - y_{j,k})}{r_{ij,k}} \qquad (5.40)$$

while G, H, p, q are the parameters of the short range forces and G* is the gravitational constant. The total force $(F_{i,k,x}, F_{i,k,y})$ on P_i due to all the other N - 1 particles is given by

$$F_{i,k,x} = \sum_{\substack{j=1 \\ j \neq i}}^{N} \overline{F}_{i,k,x}; \quad F_{i,k,y} = \sum_{\substack{j=1 \\ j \neq i}}^{N} \overline{F}_{i,k,y}. \qquad (5.41)$$

Throughout the discussion of this section, we will use $\Delta t = 10^{-4}$. In addition, a computer FORTRAN program of the basic computations to be described is given in Appendix 2.

We will consider a system of 137 particles, so that N = 137. This parameter was determined solely by economic constraints. Next, we fix the parameters q = 6, p = 4, which were found to be viable in preliminary computations with relatively smaller particle sets. Unless indicated otherwise, initial particle positions will always be those shown in Fig. 5.10 and listed precisely in the x_0, y_0 columns of Table 5.3. These positions were generated in such a fashion that the configuration is approximately circular and is, for zero initial velocities and constant masses, relatively stable.

The entire configuration will be set into counterclockwise rotation as follows. In terms of the angular velocity parameter $\dot\theta$ and a perturbation parameter ε, let

$$v_{i,0,x} = \pm |\dot\theta y_{i,0}| \pm \frac{m_i}{2000} \varepsilon, \quad v_{i,0,y} = \pm |\dot\theta x_{i,0}| \pm \frac{m_i}{2000} \varepsilon, \qquad (5.42)$$

where the choices of the signs will be made as follows. Choose the signs before the absolute value terms in (5.42) by setting $\varepsilon = 0$ and using the rule

$(x_{i,0}, y_{i,0}) \in$ Quadrant I $\Rightarrow v_{i,0,x} \leq 0, \quad v_{i,0,y} \geq 0$

$(x_{i,0}, y_{i,0}) \in$ Quadrant II $\Rightarrow v_{i,0,x} \leq 0, \quad v_{i,0,y} \leq 0$

$(x_{i,0}, y_{i,0}) \in$ Quadrant III $\Rightarrow v_{i,0,x} \geq 0, \quad v_{i,0,y} \leq 0$

$(x_{i,0}, y_{i,0}) \in$ Quadrant IV $\Rightarrow v_{i,0,x} \geq 0, \quad v_{i,0,y} \geq 0.$

As will be effected later, the signs before the ε terms will be determined at random.

Still another parameter which will be important will be a distance parameter D which will determine whether the long range forces or the short range forces predominate. We will choose D > 1 and will use the "switching" rule

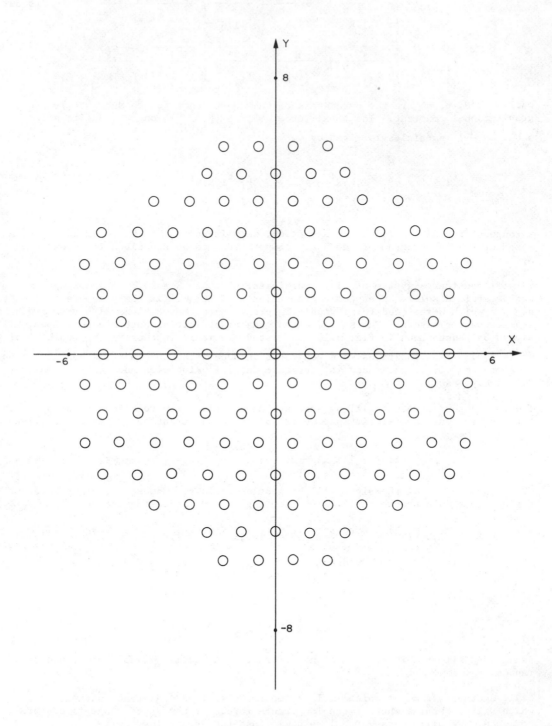

Fig. 5.10

TABLE 5.3 Initial Conditions

i	m_i	$x_{i,0}$	$y_{i,0}$	$v_{i,0,x}$	$v_{i,0,y}$
1	10000.000	-2.013	3.447	-63.700	-58.093
2	10000.000	0.487	2.590	-60.548	-48.656
3	8000.000	-4.489	-2.614	50.506	-57.819
4	8000.000	1.496	0.874	-36.596	34.207
5	8000.000	-2.503	2.595	-29.911	-50.305
6	6000.000	2.005	-5.198	49.930	22.043
7	6000.000	-0.991	-1.736	36.643	-34.130
8	6000.000	5.497	-0.869	-33.072	8.022
9	6000.000	-5.494	0.858	26.435	-52.036
10	6000.000	-1.002	1.727	-23.062	-32.590
11	6000.000	0.989	5.197	- 8.967	-26.022
12	6000.000	-1.991	-3.469	44.047	-38.052
13	6000.000	4.504	-2.599	39.992	11.172
14	6000.000	3.499	4.302	-11.928	16.192
15	6000.000	2.002	-1.726	23.000	36.942
16	6000.000	-3.004	3.448	-42.176	-17.779
17	6000.000	4.991	1.742	-36.756	50.254
18	4000.000	-3.499	2.601	10.161	6.379
19	4000.000	-0.498	-4.334	- 2.860	-21.963
20	4000.000	4.996	0.002	-19.979	0.281
21	4000.000	-1.498	-4.337	- 2.833	-26.232
22	4000.000	-5.507	2.590	-30.227	-42.401
23	4000.000	-3.001	-1.737	-13.200	-32.448
24	4000.000	0.510	-6.064	44.571	-17.775
25	4000.000	-1.990	-5.207	40.949	-27.638
26	4000.000	5.007	-3.458	33.898	- 0.198
27	4000.000	1.007	-3.464	33.939	-16.214
28	4000.000	3.007	-1.731	27.145	- 8.222
29	4000.000	-2.004	0.001	-19.751	11.913
30	4000.000	4.496	0.880	-23.345	37.608
31	4000.000	1.493	2.607	-30.421	25.221
32	4000.000	4.993	3.470	-34.039	39.965
33	2000.000	1.508	-6.055	34.193	15.821
34	2000.000	0.008	-5.198	30.864	10.301
35	2000.000	-0.992	-5.198	30.465	5.822
36	2000.000	0.508	-4.328	27.603	11.559
37	2000.000	1.508	-4.325	27.164	15.831
38	2000.000	3.505	-2.595	20.310	24.138
39	2000.000	-3.495	-2.601	20.675	- 4.072
40	2000.000	0.005	-1.728	17.189	9.776
41	2000.000	3.502	-0.865	13.802	23.993
42	2000.000	-2.498	-0.871	13.871	- 0.112
43	2000.000	3.002	0.005	9.612	22.053
44	2000.000	-1.498	0.869	6.335	4.096
45	2000.000	-0.498	0.872	6.963	8.070
46	2000.000	-2.498	0.869	6.860	0.649
47	2000.000	0.999	1.732	2.741	14.224
48	2000.000	-5.001	1.726	3.227	- 9.785
49	2000.000	-4.001	1.726	3.055	- 5.651
50	2000.000	-0.001	3.462	- 4.250	10.021
51	2000.000	0.496	4.332	- 7.248	11.301
52	2000.000	-1.504	4.329	- 7.790	3.488
53	2000.000	1.992	5.201	-30.436	- 2.226

Arithmetic Applied Mathematics

TABLE 5.3   (Continued)

i	m_i	$x_{i,0}$	$y_{i,0}$	$v_{i,0,x}$	$v_{i,0,y}$
54	2000.000	-0.508	6.058	-34.086	-11.939
55	2000.000	1.004	-5.202	10.675	- 5.577
56	2000.000	2.504	-4.329	7.105	0.065
57	2000.000	0.001	-3.462	3.390	- 9.927
58	2000.000	-2.499	-2.605	0.397	-19.737
59	2000.000	-4.999	-1.738	- 3.122	-30.063
60	2000.000	4.498	-0.866	- 6.699	7.907
61	2000.000	-1.502	-0.875	- 6.675	-15.910
62	2000.000	3.998	0.004	- 9.577	5.837
63	2000.000	0.498	0.868	-13.788	- 7.132
64	2000.000	-4.502	0.862	-13.455	-27.916
65	2000.000	-3.502	0.865	-13.429	-23.805
66	2000.000	-0.005	1.728	-16.122	-10.123
67	2000.000	-4.505	2.592	-19.992	-27.953
68	2000.000	0.995	3.458	-23.847	- 5.399
69	2000.000	-1.005	3.458	-23.940	-14.888
70	2000.000	1.492	4.331	-27.334	- 4.517
71	2000.000	-0.508	4.328	-27.502	-12.539
72	2000.000	-1.508	6.055	-34.261	-15.724
73	2000.000	3.504	-4.329	7.267	3.980
74	2000.000	-2.496	-4.335	7.436	-20.447
75	2000.000	4.001	-3.456	3.750	5.980
76	2000.000	-0.999	-3.462	3.939	-13.799
77	2000.000	-4.999	-3.468	3.800	-30.306
78	2000.000	5.501	-2.596	0.415	12.575
79	2000.000	-1.499	-2.605	0.369	-15.970
80	2000.000	1.005	-1.732	16.755	- 6.301
81	2000.000	-3.995	-1.738	16.866	-26.138
82	2000.000	-0.498	-0.872	13.320	-11.742
83	2000.000	0.502	-0.872	13.561	- 7.918
84	2000.000	-5.498	-0.878	13.698	-31.834
85	2000.000	2.002	0.001	10.032	- 2.358
86	2000.000	-3.998	-0.008	9.796	-25.846
87	2000.000	3.502	0.871	6.472	4.005
88	2000.000	5.502	0.874	6.538	11.540
89	2000.000	1.999	1.731	2.607	- 1.494
90	2000.000	5.499	2.604	- 0.471	11.759
91	2000.000	-1.501	2.595	- 0.292	-16.074
92	2000.000	1.999	3.461	- 4.224	- 1.783
93	2000.000	2.496	4.331	- 7.565	0.214
94	2000.000	-2.504	4.325	- 7.488	-20.654
95	2000.000	0.496	6.058	-14.298	- 7.829
96	2000.000	-1.492	-6.065	34.425	-15.952
97	2000.000	-3.492	-4.335	27.649	-23.920
98	2000.000	2.005	-3.459	23.911	- 2.091
99	2000.000	-3.995	-3.468	24.091	-26.184
100	2000.000	-0.495	-2.602	20.272	-12.374
101	2000.000	0.505	-2.602	20.206	- 7.595
102	2000.000	4.005	-1.726	17.254	6.058
103	2000.000	1.502	-0.869	13.649	- 3.339
104	2000.000	-4.498	-0.878	13.329	-28.419
105	2000.000	-0.998	-0.002	9.984	-14.004
106	2000.000	-4.998	-0.008	9.795	-30.092

TABLE 5.3   (Concluded)

i	m_i	$x_{i,0}$	$y_{i,0}$	$v_{i,0,x}$	$v_{i,0,x}$
107	2000.000	2.502	0.871	6.578	0.041
108	2000.000	2.999	1.731	2.686	1.761
109	2000.000	−3.001	1.725	3.732	−21.165
110	2000.000	3.499	2.601	− 0.030	4.079
111	2000.000	2.495	2.605	−20.882	19.700
112	2000.000	−0.505	2.602	−21.368	8.316
113	2000.000	2.995	3.465	−23.897	21.952
114	2000.000	−4.005	3.456	−23.030	− 6.161
115	2000.000	−3.508	4.329	−26.368	− 4.922
116	2000.000	−2.008	5.199	−30.894	1.100
117	2000.000	1.492	6.065	−34.222	16.098
118	2000.000	−0.496	−6.058	13.606	8.178
119	2000.000	3.001	−3.455	3.874	22.206
120	2000.000	−2.999	−3.461	3.481	− 1.801
121	2000.000	1.501	−2.595	0.616	16.211
122	2000.000	2.501	−2.595	0.397	20.005
123	2000.000	−5.499	−2.604	0.373	−11.792
124	2000.000	5.001	−1.722	− 3.288	29.701
125	2000.000	−1.999	−1.731	− 3.065	1.824
126	2000.000	2.498	−0.865	− 6.705	20.480
127	2000.000	−3.502	−0.871	− 6.532	− 4.177
128	2000.000	−0.002	0.002	− 9.843	10.323
129	2000.000	0.998	0.002	− 9.866	13.895
130	2000.000	−3.002	−0.001	− 9.997	− 2.150
131	2000.000	3.995	1.738	−16.825	26.147
132	2000.000	−2.005	1.729	−17.136	3.032
133	2000.000	4.495	2.608	−20.408	27.820
134	2000.000	3.995	3.468	−23.657	25.962
135	2000.000	−5.005	3.456	−23.512	− 9.969
136	2000.000	−1.008	5.202	−30.861	5.542
137	2000.000	−0.008	5.202	−30.831	9.907

$$r_{ij,k} < D \Rightarrow G* = 0 \tag{5.43}$$

$$r_{ij,k} \geq D \Rightarrow G = H = 0. \tag{5.44}$$

Further, as was determined to be reasonable in preliminary calculations, we will fix $G = H = 5$ whenever (5.43) is valid. Thus, once D is assigned, G and H are completely determined, but G* is, as yet, still arbitrary whenever (5.44) is valid.

In the assignment of particle masses, let us first consider the chemical composition of Earth, as shown in Table 5.4. The proportionate total mass for each element is shown in the right-most column. If one regroups all the elements in Table 5.4 as shown in Table 5.5, then one observes that, in order, the total masses of groups I – V decrease, while the individual particle masses increase. It is this observation which is incorporated approximately in our model as follows. We will consider five groups, consisting of 2, 3, 12, 15, and 105 particles, respectively, with individual particle masses of 10000, 8000, 6000, 4000, and 2000 units,

TABLE 5.4 Chemical Composition of Earth

Element	Relative number of atoms	Element's atomic weight	Product
Hydrogen	40000.0	1.008	40320
Helium	3100.0	4.003	12409
Carbon	3.5	12.01	42
Nitrogen	6.6	14.008	92
Oxygen	21.5	16.000	344
Neon	8.6	20.183	174
Sodium	0.04	22.997	1
Magnesium	0.91	24.31	22
Aluminum	0.09	26.98	2
Silicon	1.0	28.09	28
Phosphorous	0.01	30.07	0
Sulfur	0.37	32.06	12
Argon	0.15	39.948	6
Calcium	0.05	40.08	2
Iron	0.6	55.85	34
Nickel	0.03	58.71	2

TABLE 5.5 Regrouping of Earth's Elements

Group	Elements	Total Products
I	Hydrogen, Helium	52729
II	Carbon through Oxygen	478
III	Neon through Aluminum	199
IV	Silicon through Calcium	48
V	Iron, Nickel	36

respectively. The actual assignment of a mass to each particle will be implemented
later by a random process.

Next, while the system is in rotation, we will require a rule for determining the
physical state of each particle. Such matters, as indicated in Section 5.4, are
exceedingly complex [5], and, at present, we will develop an approach which is
relatively more sophisticated than the previous one. Consider four particles P_1,
P_2, P_3, P_4, each of the same mass, located at (0,1), (-0.87, 0.5), (0.87, 0.5),
(0,0), respectively. If each particle is assigned $\vec{0}$ velocity, then the three
particles P_1, P_2, P_4 form a three particle bond since, by (5.39)-(5.41), the
local force on any one due to the other two is zero. Similarly, P_1, P_3, P_4 form
a three particle bond. We will then call P_1, P_2, P_3 and P_4 <u>solid</u> particles. To
explore the change of state to fluid particles, we now change only the initial
velocity of P_1 to (0,v), where v > 0, and seek the smallest value of v for
which P_1 relocates monotonically upward, in accordance with (5.39)-(5.41), so
that the given bonds are broken and new ones formed by the triplet P_1, P_2, P_3 and
by the triplet P_2, P_3, P_4. In this fashion, the speed v assigned to P_1 has
enabled the particles to change their bonds easily, which we call a fluid state,
and, in particular, a liquid state. For D = 2.1, so that (5.43) is valid, these
values of v are given in Table 5.6 in the "v-liquid" column for the different
masses to be considered. (Of course, these results are also valid for any D > 2.1).
However, it will be more convenient to identify a particle as being a liquid particle by its temperature, which was defined in Section 4.4 for flow of heat in a bar,
but will now be restated using a normalized formula which makes the readings more
convenient for the present problems. The instantaneous temperature $T'_{i,k}$ of P_i
at t_k is defined by its kinetic energy, that is

$$T'_{i,k} = \frac{1}{2} m_i v^2_{i,k} . \tag{5.45}$$

Since these numbers can be relatively large due to the magnitudes m_i, the normalized instantaneous temperature $T^*_{i,k}$ of P_i at t_k is defined by

$$T^*_{i,k} = T'_{i,k}/10^4 .$$

TABLE 5.6 Change of State Velocities and Temperatures

m_i	v-liquid	v-gas	temp-liquid	temp-gas
10000	100	170	11370	30200
8000	90	160	8190	22600
6000	78	140	4640	12500
4000	65	110	1160	5300
2000	50	80	710	2130

The temperature $T_{i,k}$ of P_i at t_k is defined by

$$T_{i,k} = \frac{1}{M} \sum_{j=k-M}^{k} T^*_{i,j} , \tag{5.46}$$

where M is a positive integer, and where (5.46) is an average over M time steps. Computations have shown that it is reasonable in the examples which follow to choose $M = 500$, so that we will use

$$T_{i,k} = \frac{1}{500} \sum_{j=k-500}^{k} T^*_{i,j} . \tag{5.47}$$

The temperatures at which each particle changes state from solid to liquid are listed in the "temp-liquid" column of Table 5.6.

The critical velocities and temperatures of gas particles, as shown in Table 5.6, were determined more simply by considering only three particles P_1, P_2, P_3, located at $(0, 0.87)$, $(-0.5, 0)$, $(0.5, 0)$, respectively, with initial velocities $(0,v)$, $(0,0)$, $(0,0)$, respectively, and by determining the positive parameter v for which $|P_1P_2| = |P_1P_3| > D$. The results shown are for $D = 2.3$. Intuitively, when $|P_1P_2| > D$, all molecular-type forces are zero, so that P_1 moves "freely", which is characteristic of gas particles.

Note that "temperature", as defined above, is a phenomenon of a particle's "local" velocity, that is, its velocity relative to neighboring particles. Thus, when a particle is rotating within a large system, which will be the case in the examples which follow, the gross system velocities should have no effect on the particle's temperature and must be subtracted out before the temperature calculation is performed. The velocity of the centroid of the system and of the average angular velocity of the system are utilized for this purpose in the following way to determine the temperature of P_i at t_k as P_i rotates within the system. First, at time t_k, let the mass center of the system be $(\bar{x}_k, \bar{y}_k)$ and let the average linear velocity, $(\bar{v}_{x,k}, \bar{v}_{y,k})$, of the system be defined by

$$\bar{v}_{x,k} = \frac{\Sigma(m_i v_{x,i,k})}{\Sigma m_i} , \quad \bar{v}_{y,k} = \frac{\Sigma(m_i v_{y,i,k})}{\Sigma m_i} , \tag{5.48}$$

where the summations of (5.48) are taken over all particles of the system. Then P_i's position $(x^*_{i,k}, y^*_{i,k})$ and velocity $(v^*_{i,k,x}, v^*_{i,k,y})$ relative to the mass center are defined by

$$x^*_{i,k} = x_{i,k} - \bar{x}_k, \quad y^*_{i,k} = y_{i,k} - \bar{y}_k \tag{5.49}$$

$$v^*_{i,k,x} = v_{i,k,x} - \bar{v}_{x,k}, \quad v^*_{i,k,y} = v_{i,k,y} - \bar{v}_{y,k}. \tag{5.50}$$

Next, out of $v^*_{i,k,x}$ and $v^*_{i,k,y}$ we wish to take the angular rotation of the system, which is done as follows. Introduce the normal and tangent velocity components, $v^*_{i,k,n}$ and $v^*_{i,k,t}$, respectively, of P_i at t_k, by the usual formulas

$$v^*_{i,k,n} = [v^*_{i,k,y} y^*_{i,k} + v^*_{i,k,x} x^*_{i,k}]/R_i \qquad (5.51)$$

$$v^*_{i,k,t} = [-v^*_{i,k,x} y^*_{i,k} + v^*_{i,k,y} x^*_{i,k}]/R_i, \qquad (5.52)$$

where

$$R_i = [(x^*_{i,k})^2 + (y^*_{i,k})^2]^{\frac{1}{2}}. \qquad (5.53)$$

Since, in general, $\dot{\theta} = v_t/R$, we define the average angular velocity, $\overline{\dot{\theta}}$, of the system by

$$\overline{\dot{\theta}} = \frac{\sum(\dot{\theta}_i m_i)}{\sum m_i} = \frac{\sum(m_i \dfrac{v^*_{i,k,t}}{R_i})}{\sum m_i}, \qquad (5.54)$$

where the summations are taken over all the particles of the system. Finally, the speeds $v^2_{i,k}$, used in (5.45) to calculate the temperature by (5.47), are given by

$$v^2_{i,k} = (v^*_{i,k,t} - \overline{\dot{\theta}} R_i)^2 + (v^*_{i,k,n})^2. \qquad (5.55)$$

As a last consideration, we will allow for radiation into and out of the system. This will be accomplished as follows. Consider the vertical strip regions

$$k\gamma \le x \le (k+1)\gamma; \quad \gamma > 0, \quad k = 0, \pm 1, \pm 2, \pm 3, \dots .$$

In each strip, the particles with maximum and minimum y coordinates are called outer particles. An outer particle whose $y_{i,k}$ is a maximum in a given strip will be called a _light-side_ particle, that is, it faces a sun. An outer particle whose $y_{i,k}$ is a minimum in a given strip is called a _dark-side_ particle. All light-side particles will receive radiant heat, while all dark-side particles will emanate radiant heat. This is implemented as follows. Every k^* steps, outer particle velocities are reset by the rule

$$\text{light-side particle:} \quad \vec{v}_{i,k} \rightarrow 1.001 \, \vec{v}_{i,k}$$

$$\text{dark-side particle:} \quad \vec{v}_{i,k} \rightarrow 0.9955 \, \vec{v}_{i,k} . \qquad (5.56)$$

A variety of examples were then run with various combinations of parameter choices selected from $\dot{\theta} = 4, 5$; $\varepsilon = 5, 10$; $k^* = 50, 100$; $D = 2.1, 2.3$; $\gamma = 0.1, 1.0, 2.0$; and $G^* = 0.01, 0.001$. From these we will first describe two, in both of which the parameter choices are $\dot{\theta} = 4$, $\varepsilon = 10$, $D = 2.3$, $k^* = 100$, $\gamma = 1$, and $G^* = 0.001$. Later, some general remarks will be made about other computations. Initial positions and velocities were generated in the manner described for (5.42) and, for completeness, are given in Table 5.3.

Example 1. In this example, we consider the mass distribution given in Table 5.3. This distribution was generated at random, but under the constraint that the relatively heavy particles be more centrally located than the relatively light ones. Such a constraint is reasonable because heavy particles on the periphery of a rotating system often have sufficient momentum to escape the system, which was

verified computationally for the present system. Thus, in effect, we are starting
with a system from which some of the outer particles which will escape are considered to have already escaped.

The rotational motion of the system, relative to the centroid, is shown in Figures
5.11(a-j) at the respective times t_{10000}, t_{15000}, t_{20000}, t_{25000}, t_{30000}, t_{35000},
t_{40000}, t_{45000}, t_{50000}, and t_{60000}. The five different size circles, from largest
to smallest, represent particles from the five sets of masses, from largest to
smallest, respectively. For emphasis, the interior of the circle representing any
particle of mass greater than 2000 has been darkened and this darkened set will
be referred to as the "heavier particles". When any particle's distance to the
centroid exceeded 15 units, that particle was considered to have escaped the system and it was dropped from all further considerations.

The system's self-reorganization with the heavier particles located centrally is
clear from the figures. Figures 5.11(a-c) indicate that the process occurs in two
steps: first the heavier particles form into several clusters, and then these
clusters relocate centrally. By the time t_{30000}, the system has been reduced to
sixty four particles and has achieved a state of relative stability. It is of
interest to note that <u>all</u> the particles which have escaped from the system have
mass 2000. At time t_{60000}, the geometric center of the system is (-0.06,-0.09),
which is almost, but not quite, identical with the centroid. It is interesting to
note that at this time the two heaviest particles are <u>not</u> located at the centroid.
Such anomalies with regard to the centroid are often present in lunar type bodies.
To explain such phenomena, let us examine Figs. 5.12(a-j), which show the state of
solidification at the very same times as those in Figs. 5.11(a-j). Using Table
5.6, we have shown the solid particles by means of hexagonal enclosures and the gas
particles by means of triangular enclosures. All other particles are liquid. If
one studies the two heaviest particles in Fig. 5.12(c) and observes the relative
positions of the neighboring solid particles, then it is apparent that these particles have already formed into a rigid, noncircular configuration which does not
change during the remainder of the motion. It is this very early solidification
which prevents those inner motions which are necessary for the heaviest particles
to relocate to the centroid of the entire system.

Figure 5.13 shows the oscillatory dissipation of system kinetic energy until time
t_{60000}, at which time the system is relatively stable.

Example 2. In order to avoid the early solidification process discussed in Example
1, the following nine changes in mass were made: $m_1 = m_3 = m_5 = m_6 = m_8 = m_9 = m_{11}$
= m_{17} = 4000, m_7 = 8000. All other considerations were identical and the system
was again studied through t_{60000}. The resulting motion is shown in Figs. 5.14
(a-i) at the respective times t_{10000}, t_{15000}, t_{20000}, t_{30000}, t_{35000}, t_{40000},
t_{45000}, t_{50000}, t_{60000}. The increase in fluidity has resulted by t_{20000} in a
separation into two distinct subsystems. It is the larger subsystem, shown to the
left in Fig. 5.14(c) which is then studied through t_{60000}. At t_{30000} this system has 52 particles, as shown in Fig. 5.14(d). By the time t_{60000}, the largest particle is at the center of the system, the geometric center is (-0.004,
-0.02), and the system is almost circular. The density clearly decreases monotonically from the center outward. The solidification process is shown in Fig.
5.15(a-c) and Fig. 5.15(c), at time t_{60000}, is certainly consistent with current
theories on the structure of the interiors of planets.

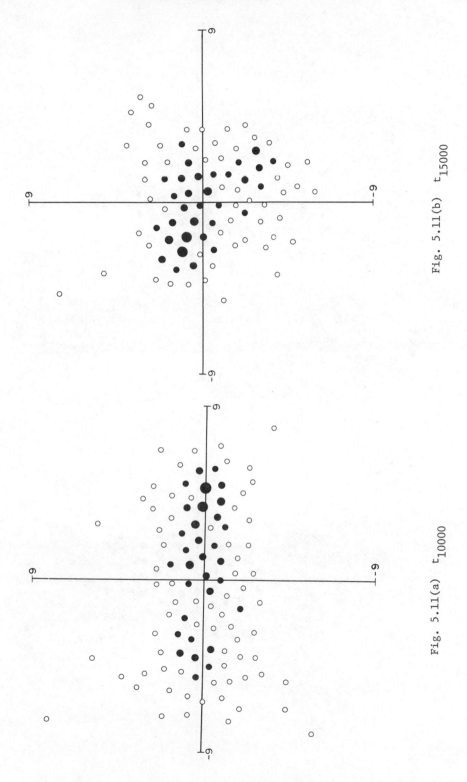

Fig. 5.11(b) t_{15000}

Fig. 5.11(a) t_{10000}

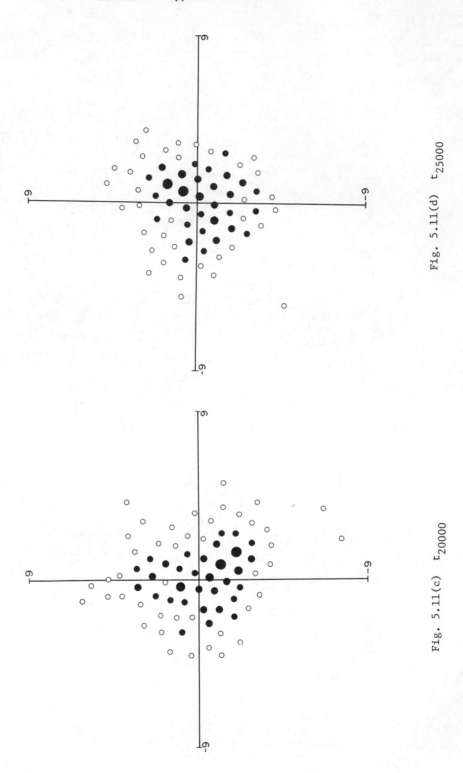

Fig. 5.11(d) t_{25000}

Fig. 5.11(c) t_{20000}

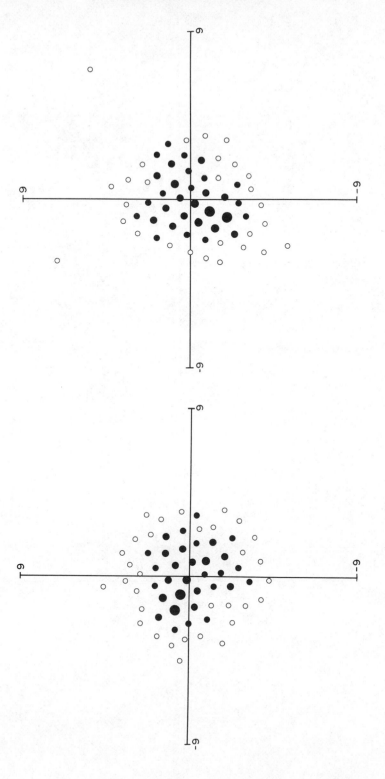

Fig. 5.11(f) t_{35000}

Fig. 5.11(e) t_{30000}

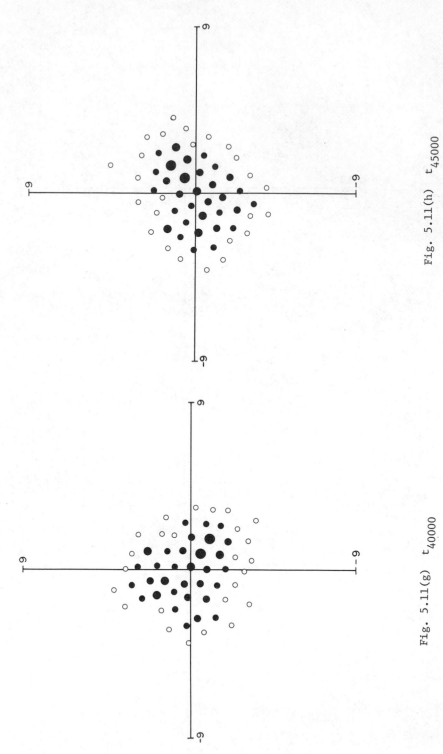

Fig. 5.11(h) t_{45000}

Fig. 5.11(g) t_{40000}

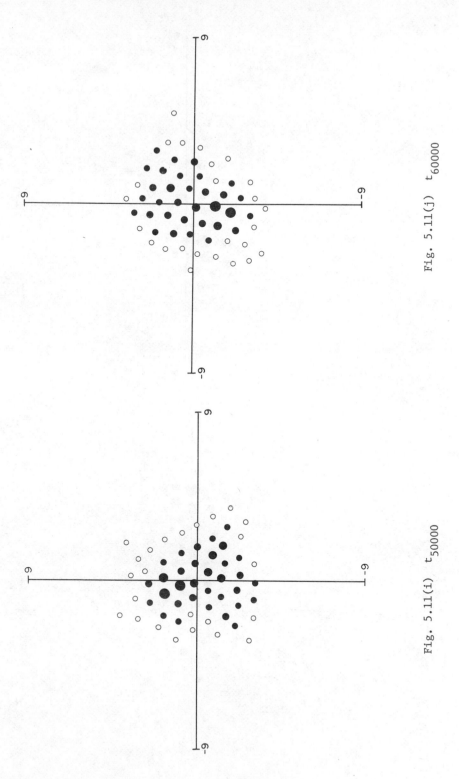

Fig. 5.11(j) t_{60000}

Fig. 5.11(i) t_{50000}

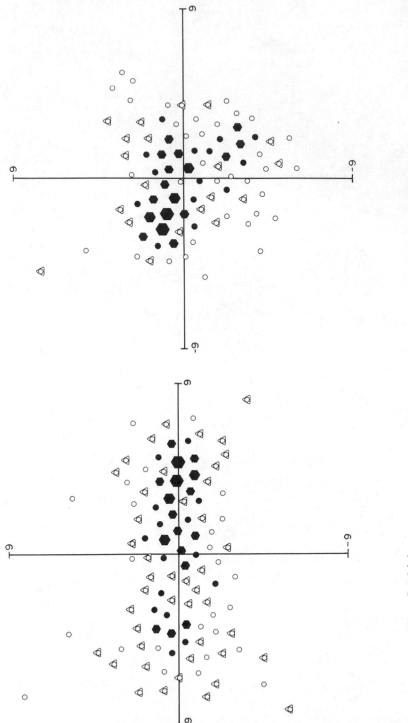

Fig. 5.12(b) t_{15000}

Fig. 5.12(a) t_{10000}

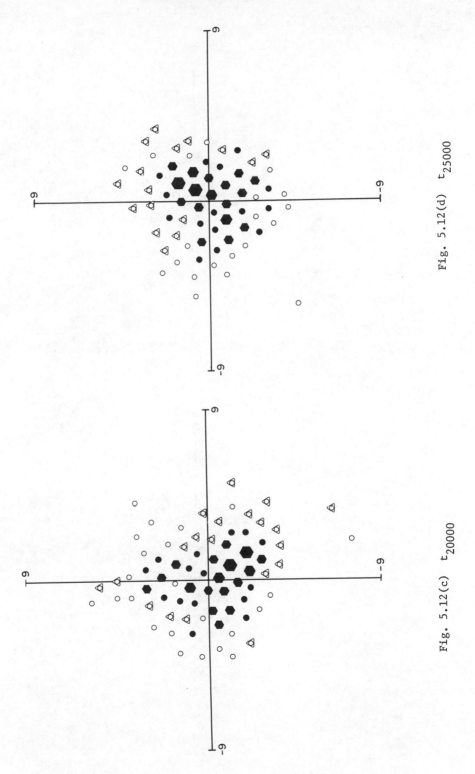

Fig. 5.12(d) t_{25000}

Fig. 5.12(c) t_{20000}

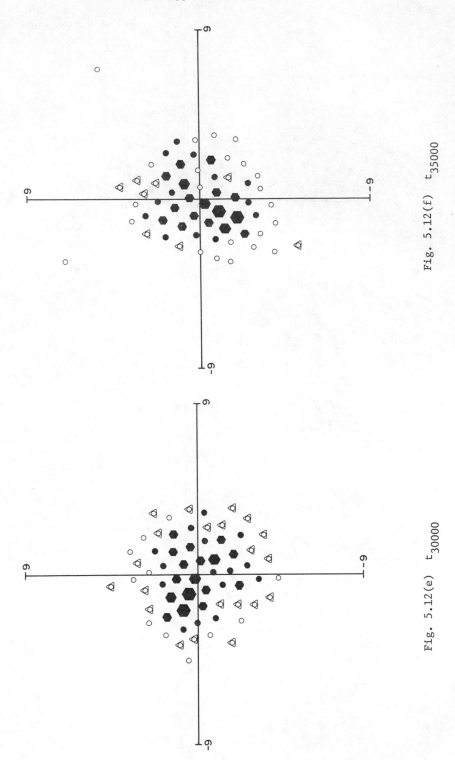

Fig. 5.12(f) t_{35000}

Fig. 5.12(e) t_{30000}

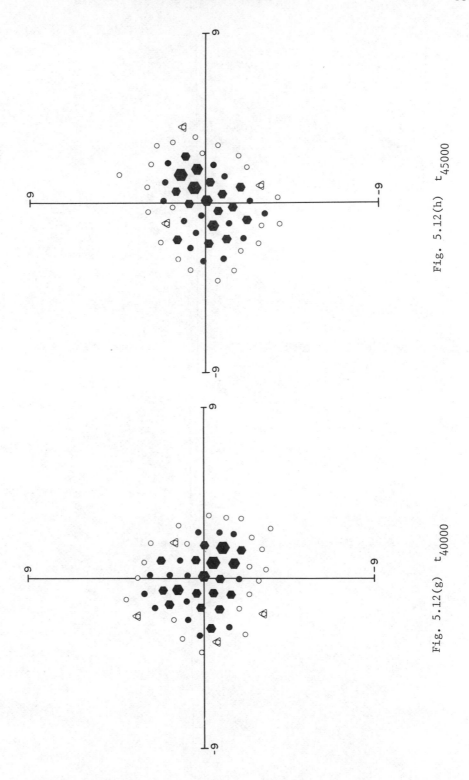

Fig. 5.12(h) t_{45000}

Fig. 5.12(g) t_{40000}

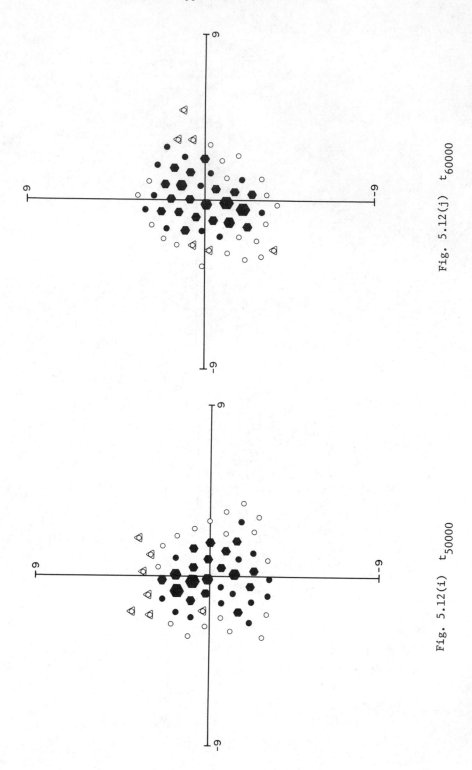

Fig. 5.12(i) t_{50000}

Fig. 5.12(j) t_{60000}

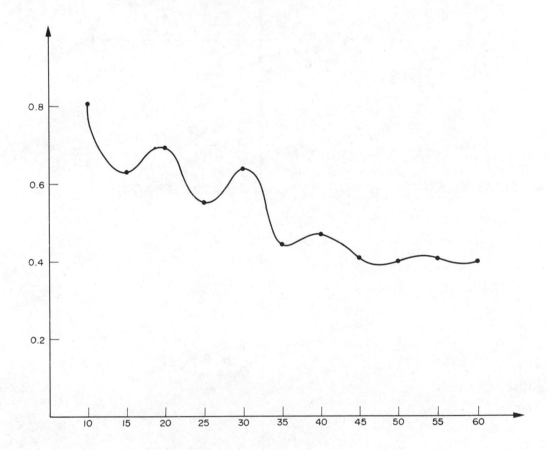

Fig. 5.13.   Kinetic energy dissipation

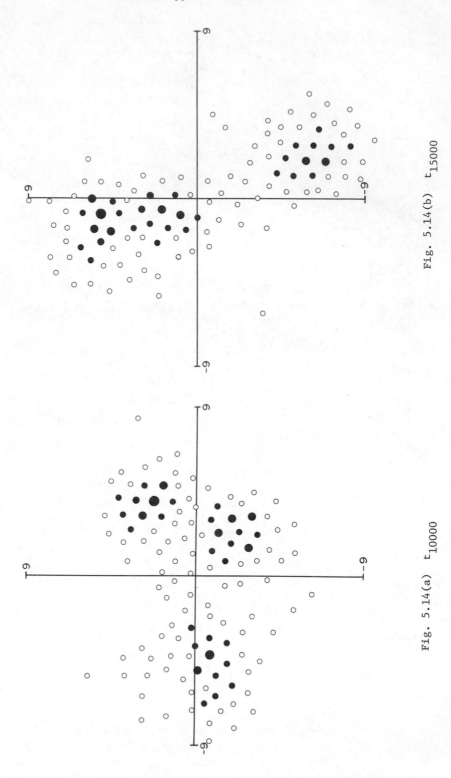

Fig. 5.14(a) t_{10000}

Fig. 5.14(b) t_{15000}

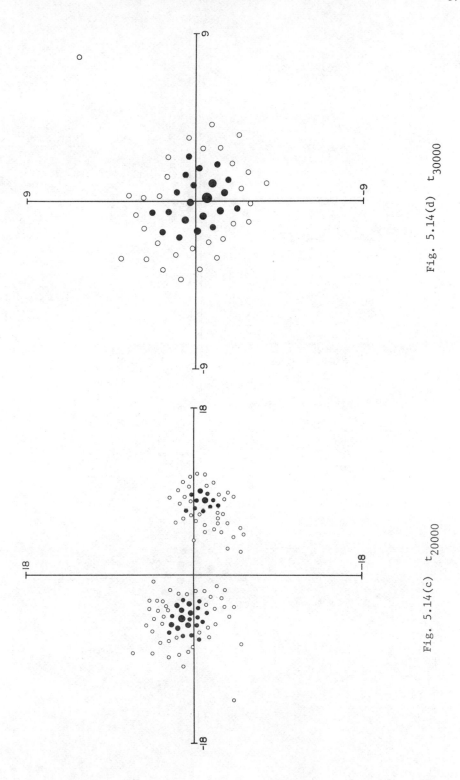

Fig. 5.14(d) t_{30000}

Fig. 5.14(c) t_{20000}

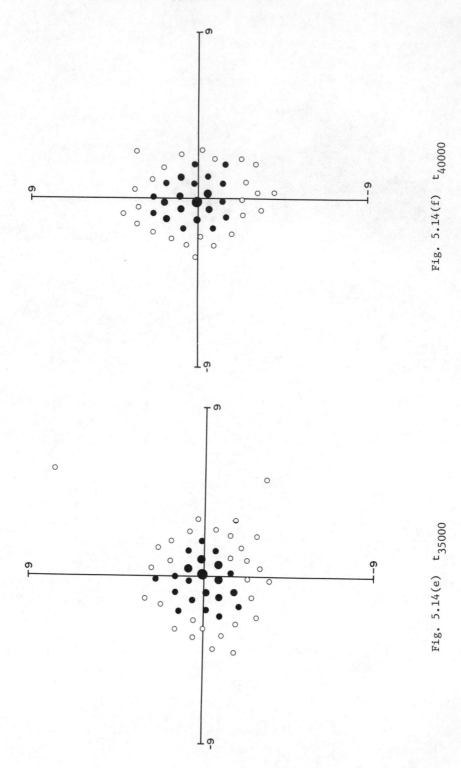

Fig. 5.14(f) t_{40000}

Fig. 5.14(e) t_{35000}

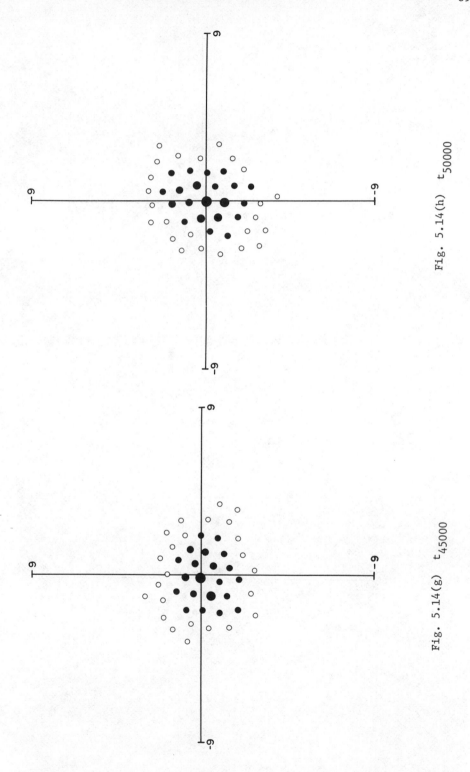

Fig. 5.14(h) t_{50000}

Fig. 5.14(g) t_{45000}

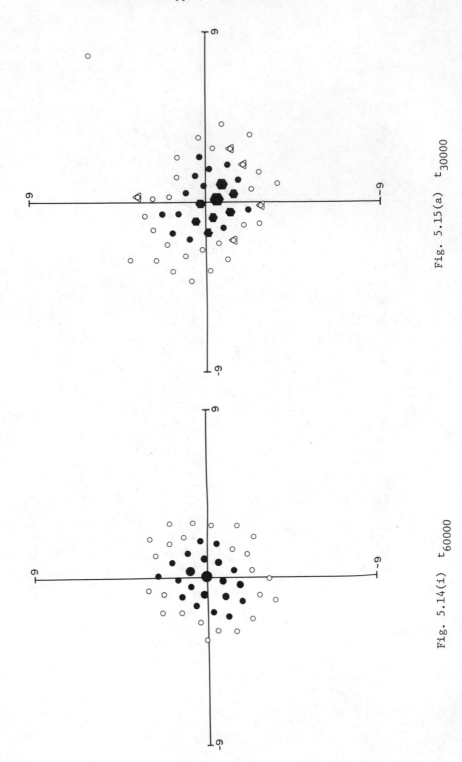

Fig. 5.15(a) t_{30000}

Fig. 5.14(i) t_{60000}

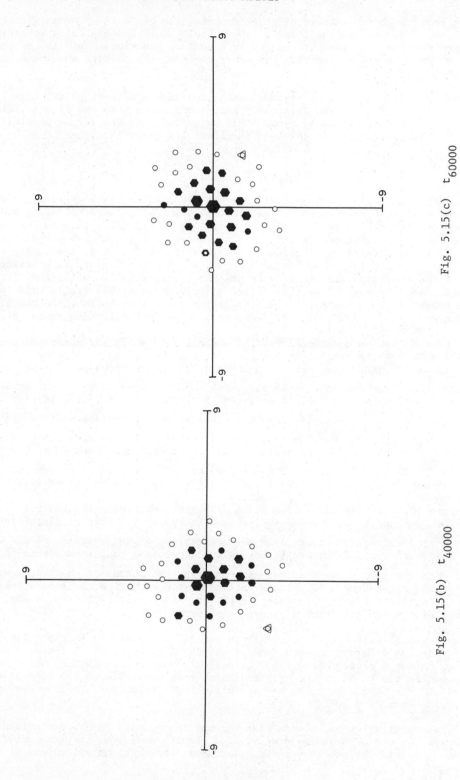

Fig. 5.15(c) t_{60000}

Fig. 5.15(b) t_{40000}

The results of the variety of other examples run can only be summarized easily as
follows. As in Examples 1 and 2, above, only minor changes in input parameters
often resulted in dramatic changes in dynamical behavior. This is, of course,
consistent with the diversity one observes among the planets and moons of the solar
system.

Other computations, aimed at duplicating the development of galaxy arms, all ended
in failure. The systems invariably self-reorganized into one or more relatively
elliptic or circular subsystems. This suggests to us the strong possibility that
more than gravitation is involved in galaxy arm formation.

5.6   FREE SURFACE FLUID FLOW

In sections 5.4 and 5.5, special care was given to the large variety of details
involved in the development of viable models which included both long and short
range forces and whose dynamical equations were explicit. In this and in the next
sections, we will continue this type of modeling, but the emphasis will be more on
physical aspects of the problems. The local force formulas, again, are (5.15) and
(5.16), and the long range force is gravity. The problems to be considered relate
to fluid phenomena and are of wide practical interest. We will study first the
motion which results from the fall of a drop into a stable liquid well (or cavity,
or reservoir). Stratified flow, the diffusion of a pollutant, and the breakwater
problem ([43], [47], [48], [77]) can be modeled in the fashion to be described.

In a well, let us consider a particle-type liquid which was generated previously
[45] in a fashion completely analogous to that described in Section 5.4. It consists of 190 particles, each of unit mass, which are shown at t_0 in Fig. 5.16.
The liquid drop consists of 15 particles, each of mass $m = 2$, and these are
represented by the shaded particles in Fig. 5.16. At t_0, the initial drop configuration assumes that the drop has already hit the surface and has flattened.
The parameters are $H = G = 100$, $p = 1$, $q = 3$, $\Delta t = 10^{-4}$, with gravity equal to
$-980m_i$. Local forces are restricted to radial distances less than 0.25 and a
velocity damping factor of 0.9 is used to reflect a particle symmetrically after
collision with a container wall [43]. All initial data are given in Table 5.7.

Figures 5.17–5.25 show the resulting dynamical behavior. Figures 5.17 and 5.18,
at t_{200} and t_{400}, respectively, show the entry and initial dispersion of the
drop into the well. Figure 5.19 shows clearly the wave reaction of the well
liquid. Figure 5.20 at t_{800}, shows two developing reactions within the well, one
of which is a backflow over the sinking drop, the second of which is a wave flow
over the right-hand wall. These actions continue in Figs. 5.21–5.25, which show,
in addition, a distinct decrease in the drop's vertical velocity, which, in the
case of a pollutant, causes a layered effect.

Figures 5.26–5.29 elucidate the gross motion of the liquid in the well by illus-
trating the deformation with time of various liquid columns. The accordian type
deformation directly below the drop suggests vortex circulation, while the column
motion to the right of the drop indicates that particles on the tops of columns
tend to flow over the right barrier while those below tend to flow over the sub-
merging drop.

Various similar examples were also run and the results can be summarized as follows. The use of a square drop yielded results analogous to those described
above, as did the use of a rectangular drop. However, in all cases, if the mass of

TABLE 5.7 Liquid Drop Data

i	m_i	x	y	v_x	v_y
1	1.00000	0.0	2.31000	0.42000	-0.90000
2	1.00000	0.0	1.98000	0.20000	-0.60000
3	1.00000	0.0	1.73000	0.0	-0.80000
4	1.00000	0.0	1.48000	0.0	-0.30000
5	1.00000	0.0	1.23000	0.10000	-0.60000
6	1.00000	0.0	0.98000	0.0	0.44000
7	1.00000	0.0	0.73000	-0.80000	-0.80000
8	1.00000	0.0	0.49000	0.0	0.0
9	1.00000	0.0	0.24000	0.0	0.0
10	1.00000	0.0	0.0	0.0	0.0
11	1.00000	0.19000	0.0	0.0	0.13000
12	1.00000	0.21000	2.11000	4.10000	-6.90000
13	1.00000	0.24000	0.19000	-1.42000	3.23000
14	1.00000	0.23000	0.38000	4.50000	4.30000
15	1.00000	0.22000	0.62000	-3.00000	4.00000
16	1.00000	0.22000	0.86000	-4.00000	4.10000
17	1.00000	0.22000	1.10000	0.60000	2.70000
18	1.00000	0.23000	1.35000	1.60000	-1.00000
19	1.00000	0.22000	1.60000	2.20000	0.20000
20	1.00000	0.22000	1.85000	-1.50000	-0.30000
21	1.00000	0.41000	0.0	0.0	0.20000
22	1.00000	0.46000	0.24000	-2.30000	6.70000
23	1.00000	0.44000	0.49000	2.10000	0.78000
24	1.00000	0.45000	0.73000	6.10000	0.75000
25	1.00000	0.44000	0.98000	0.76000	3.40000
26	1.00000	0.44000	1.23000	-2.40000	-6.10000
27	1.00000	0.44400	1.48000	2.10000	-2.40000
28	1.00000	0.44400	1.73000	0.30000	2.00000
29	1.00000	0.44000	1.98000	3.30000	-0.30000
30	1.00000	0.43200	2.24000	0.36000	-0.60000
31	1.00000	0.60500	0.0	0.0	-0.80000
32	1.00000	0.70000	0.19000	-4.30000	-2.30000
33	1.00000	0.67000	0.39000	1.60000	-1.70000
34	1.00000	0.67000	0.61000	-3.40000	1.90000
35	1.00000	0.66000	0.86000	4.76000	-1.90000
36	1.00000	0.66000	1.10000	1.20000	1.20000
37	1.00000	0.67000	1.36000	1.30000	-1.00000
38	1.00000	0.66000	1.61000	-0.66000	-2.00000
39	1.00000	0.67000	1.87000	1.70000	-6.80000
40	1.00000	0.67000	2.11000	4.30000	-1.00000
41	1.00000	0.84000	0.0	0.0	0.0
42	1.00000	0.92000	0.24000	-3.00000	2.20000
43	1.00000	0.90000	0.49000	1.90000	0.20000
44	1.00000	0.88000	0.73000	3.10000	4.00000
45	1.00000	0.88000	0.98000	-3.00000	-1.20000
46	1.00000	0.88000	1.22000	0.90000	-3.00000
47	1.00000	0.89000	1.47000	4.60000	1.20000
48	1.00000	0.88000	1.70000	2.70000	0.28000
49	1.00000	0.89000	1.98000	4.70000	-3.20000
50	1.00000	0.85000	2.29000	-0.80000	-2.00000
51	1.00000	1.04000	0.0	0.0	0.0
52	1.00000	1.16000	0.18000	4.00000	0.60000

Arithmetic Applied Mathematics

TABLE 5.7 (Continued)

i	m_i	x	y	v_x	v_y
53	1.00000	1.12000	0.38000	3.00000	−2.00000
54	1.00000	1.11000	0.61000	−5.50000	−1.60000
55	1.00000	1.11000	0.86000	−1.20000	−2.00000
56	1.00000	1.11000	1.11000	2.10000	1.00000
57	1.00000	1.11000	1.36000	4.60000	−3.40000
58	1.00000	1.11000	1.62000	−1.50000	−4.50000
59	1.00000	1.11000	1.87000	−2.80000	1.90000
60	1.00000	1.10000	2.14000	−2.50000	−5.00000
61	1.00000	1.29000	0.0	0.0	0.0
62	1.00000	1.38000	0.23000	−0.50000	−3.20000
63	1.00000	1.33000	0.49000	−2.40000	−2.00000
64	1.00000	1.33000	0.74000	6.10000	−5.00000
65	1.00000	1.33000	0.99000	3.70000	0.60000
66	1.00000	1.34000	1.24000	0.90000	2.50000
67	1.00000	1.34000	1.49000	2.00000	−4.00000
68	1.00000	1.34000	1.75000	−1.30000	−0.70000
69	1.00000	1.30000	2.00000	4.60000	2.80000
70	1.00000	1.35000	2.32000	5.30000	−6.00000
71	1.00000	1.55000	2.13000	−4.00000	0.20000
72	1.00000	1.50000	0.0	0.0	0.0
73	1.00000	1.61000	0.18000	−4.00000	4.90000
74	1.00000	1.57000	0.38000	−6.00000	0.40000
75	1.00000	1.55000	0.62000	−4.80000	2.40000
76	1.00000	1.55000	0.87000	−3.00000	5.00000
77	1.00000	1.55000	1.12000	−5.00000	0.50000
78	1.00000	1.56000	1.37000	−1.50000	1.90000
79	1.00000	1.55000	1.62000	1.50000	−0.80000
80	1.00000	1.56000	1.87000	−0.50000	−5.00000
81	1.00000	1.76000	2.26000	−2.50000	2.60000
82	1.00000	1.75000	0.0	0.0	0.0
83	1.00000	1.81000	0.26000	−3.00000	−0.70000
84	1.00000	1.77000	0.50000	−2.00000	0.10000
85	1.00000	1.76000	0.75000	−3.00000	4.80000
86	1.00000	1.77000	1.00000	−2.60000	−0.40000
87	1.00000	1.78000	1.25000	3.80000	1.70000
88	1.00000	1.78000	1.50000	1.20000	0.43000
89	1.00000	1.78000	1.76000	9.00000	−5.0000
90	1.00000	1.77000	2.00000	−4.40000	−1.30000
91	1.00000	1.98000	0.0	0.0	0.12000
92	1.00000	2.04000	0.18000	0.30000	6.00000
93	1.00000	2.00000	0.39000	5.00000	0.0
94	1.00000	1.99000	0.64000	−4.50000	−0.30000
95	1.00000	1.99000	0.88000	−1.90000	0.86000
96	1.00000	2.01000	1.13000	−2.70000	2.70000
97	1.00000	1.98000	2.14000	4.00000	−4.00000
98	1.00000	1.99000	1.38000	0.0	−2.00000
99	1.00000	1.99000	1.63000	−5.60000	−4.00000
100	1.00000	1.99000	1.89000	−2.9000	−0.40000
101	1.00000	2.22000	2.00000	3.40000	−1.80000
102	1.00000	2.22000	0.0	0.0	0.0
103	1.00000	2.22000	0.29000	3.60000	0.40000
104	1.00000	2.20000	2.29000	0.50000	−3.50000

TABLE 5.7   (Continued)

i	m_i	x	y	v_x	v_y
105	1.00000	2.20000	0.52000	−4.00000	−1.00000
106	1.00000	2.22000	0.76000	−6.00000	−7.00000
107	1.00000	2.22000	0.99000	5.00000	0.30000
108	1.00000	2.23000	1.24000	2.00000	−3.00000
109	1.00000	2.22000	1.49000	3.00000	0.10000
110	1.00000	2.23000	1.74000	7.25000	1.20000
111	1.00000	2.44000	2.15000	0.80000	2.30000
112	1.00000	2.67000	2.30000	−4.00000	3.30000
113	1.00000	2.47000	0.0	0.0	0.0
114	1.00000	2.40000	0.18000	3.10000	4.70000
115	1.00000	2.44000	0.41000	−5.50000	−3.00000
116	1.00000	2.44000	0.64000	−1.50000	6.90000
117	1.00000	2.44000	0.88000	6.60000	−2.70000
118	1.00000	2.45000	1.12000	2.60000	0.60000
119	1.00000	2.45000	1.37000	2.60000	−1.37000
120	1.00000	2.45000	1.63000	0.80000	3.50000
121	1.00000	2.44000	1.88000	6.20000	−0.20000
122	1.00000	2.66000	2.02000	−1.00000	−2.60000
123	1.00000	2.69000	0.0	−0.10000	−0.50000
124	1.00000	2.63000	0.26000	0.0	−1.10000
125	1.00000	2.67000	0.51000	1.60000	−2.40000
126	1.00000	2.66000	0.75000	5.30000	1.50000
127	1.00000	2.67000	0.99000	1.70000	2.50000
128	1.00000	2.67000	1.24000	0.0	−6.00000
129	1.00000	2.67000	1.49000	−1.80000	3.40000
130	1.00000	2.66000	1.76000	1.80000	−6.40000
131	1.00000	2.89000	2.12000	−7.00000	0.50000
132	1.00000	2.94000	0.0	0.0	0.0
133	1.00000	2.84000	0.19000	−1.00000	−1.50000
134	1.00000	2.87000	0.39000	2.10000	−5.00000
135	1.00000	2.89000	0.61000	0.90000	−5.00000
136	1.00000	2.89000	0.86000	4.51000	1.00000
137	1.00000	2.89000	1.12000	1.67000	5.60000
138	1.00000	2.89000	1.37000	−0.80000	3.30000
139	1.00000	2.89000	1.61000	1.00000	−2.00000
140	1.00000	2.89000	1.86000	−1.00000	−1.00000
141	1.00000	3.13000	2.25000	−5.00000	−1.00000
142	1.00000	3.15000	0.0	0.10000	−1.00000
143	1.00000	3.08000	0.24000	4.40000	−4.00000
144	1.00000	3.10000	0.50000	3.80000	−5.00000
145	1.00000	3.11000	0.74000	4.00000	0.10000
146	1.00000	3.11000	0.99000	0.95000	−3.00000
147	1.00000	3.12000	1.24000	0.30000	−2.00000
148	1.00000	3.11000	1.49000	−1.00000	−0.20000
149	1.00000	3.12000	1.74000	3.70000	1.99000
150	1.00000	3.12000	1.99000	−8.00000	−0.60000
151	1.00000	3.30000	0.18000	2.00000	1.40000
152	1.00000	3.32000	0.40000	0.0	3.00000
153	1.00000	3.33000	0.62000	0.70000	1.70000
154	1.00000	3.33000	0.86000	−7.00000	0.70000
155	1.00000	3.33000	1.11000	1.50000	1.00000
156	1.00000	3.34000	1.36000	6.60000	6.00000
157	1.00000	3.33000	1.61000	−1.60000	0.10000

Arithmetic Applied Mathematics

TABLE 5.7 (Concluded)

i	m_i	x	y	v_x	v_y
158	1.00000	3.34000	1.86000	2.10000	−0.40000
159	1.00000	3.39000	0.0	0.0	0.0
160	1.00000	3.34000	2.12000	1.00000	−6.00000
161	1.00000	3.56000	1.98000	0.20000	−0.30000
162	1.00000	3.59000	0.0	0.0	0.0
163	1.00000	3.52000	0.25000	0.90000	2.70000
164	1.00000	3.55000	0.49000	5.00000	−7.00000
165	1.00000	3.55000	0.73000	1.60000	1.80000
166	1.00000	3.55000	0.98000	−1.30000	−2.40000
167	1.00000	3.56000	1.23000	0.10000	−1.00000
168	1.00000	3.56000	1.48000	0.0	−0.50000
169	1.00000	3.56000	1.73000	−2.40000	−2.00000
170	1.00000	3.77000	2.12000	−2.50000	1.60000
171	1.00000	3.57000	2.25000	3.00000	−1.40000
172	1.00000	3.81000	0.0	0.0	0.0
173	1.00000	3.76000	0.19000	0.0	−0.20000
174	1.00000	3.78000	0.38000	−4.00000	0.70000
175	1.00000	3.78000	0.61000	3.10000	−0.40000
176	1.00000	3.78000	0.86000	−2.00000	−4.00000
177	1.00000	3.78000	1.11000	3.80000	−5.00000
178	1.00000	3.78000	1.36000	1.70000	−2.70000
179	1.00000	3.78000	1.61000	−4.50000	−1.00000
180	1.00000	3.78000	1.86000	0.60000	0.90000
181	1.00000	3.99000	0.0	0.0	0.0
182	1.00000	4.00000	0.24000	0.0	0.0
183	1.00000	4.00000	0.49000	0.0	0.0
184	1.00000	3.99000	0.73000	0.0	0.0
185	1.00000	4.00000	0.98000	0.0	0.0
186	1.00000	4.00000	1.23000	0.0	0.0
187	1.00000	4.00000	1.48000	0.0	0.0
188	1.00000	4.00000	1.73000	0.0	0.0
189	1.00000	4.00000	1.98000	0.0	0.0
190	1.00000	4.00000	2.23000	0.0	0.0
191	2.00000	0.0	2.50000	0.0	−50.00000
192	2.00000	0.25000	2.50000	0.0	−50.00000
193	2.00000	0.50000	2.50000	0.0	−50.00000
194	2.00000	0.75000	2.50000	0.0	−50.00000
195	2.00000	1.00000	2.50000	0.0	−50.00000
196	2.00000	0.12500	2.70000	0.0	−50.00000
197	2.00000	0.37500	2.70000	0.0	−50.00000
198	2.00000	0.62500	2.70000	0.0	−50.00000
199	2.00000	0.87500	2.70000	0.0	−50.00000
200	2.00000	0.25000	2.90000	0.0	−50.00000
201	2.00000	0.50000	2.90000	0.0	−50.00000
202	2.00000	0.75000	2.90000	0.0	−50.00000
203	2.00000	0.37500	3.10000	0.0	−50.00000
204	2.00000	0.62500	3.10000	0.0	−50.00000
205	2.00000	0.50000	3.60000	0.0	−50.00000

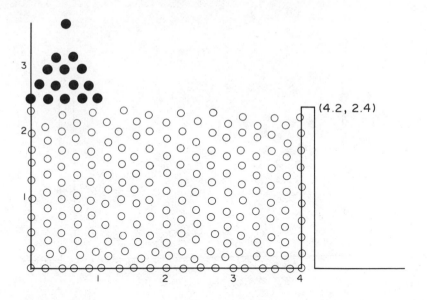

Fig. 5.16
t_0

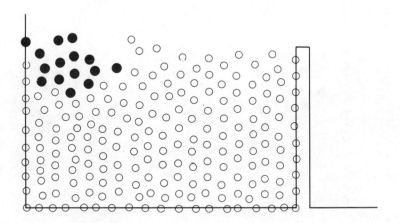

Fig. 5.17
t_{200}

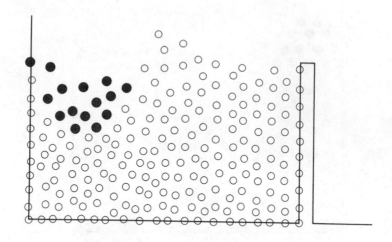

Fig. 5.18
t_{400}

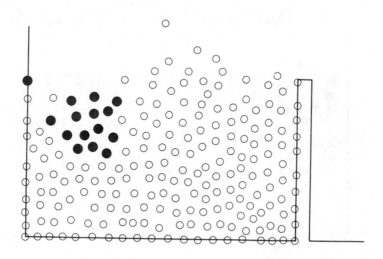

Fig. 5.19
t_{600}

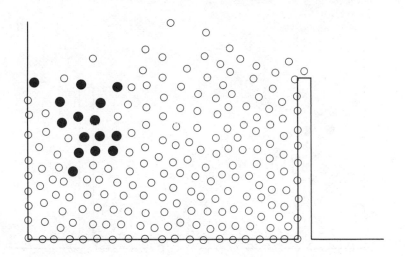

Fig. 5.20
t_{800}

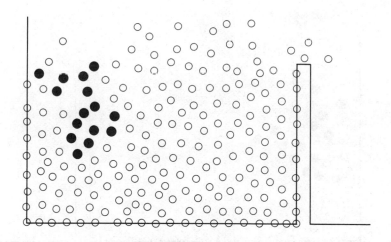

Fig. 5.21
t_{1000}

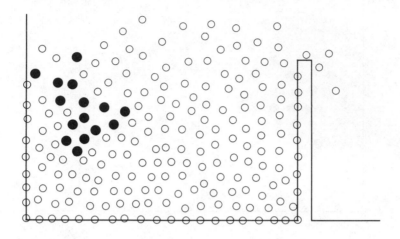

Fig. 5.22
t_{1200}

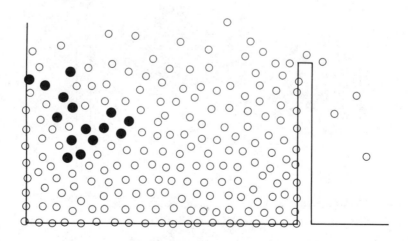

Fig. 5.23
t_{1400}

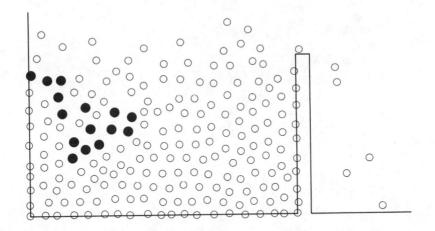

Fig. 5.24
t_{1600}

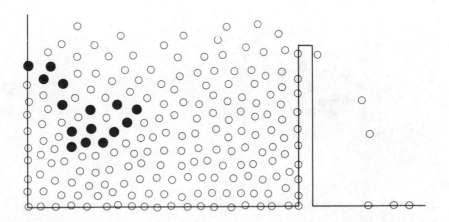

Fig. 5.25
t_{1800}

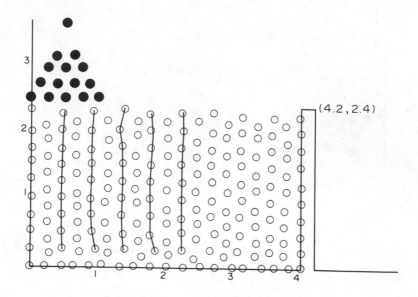

Fig. 5.26
t_0

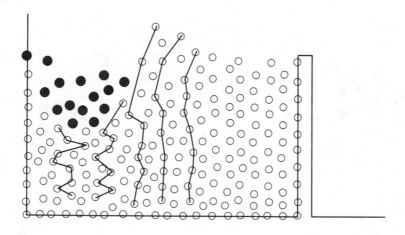

Fig. 5.27
t_{400}

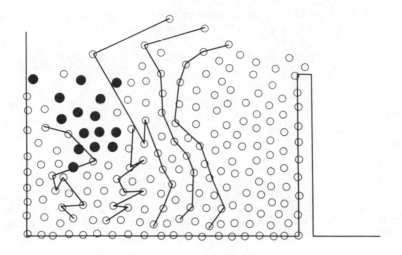

Fig. 5.28
t_{800}

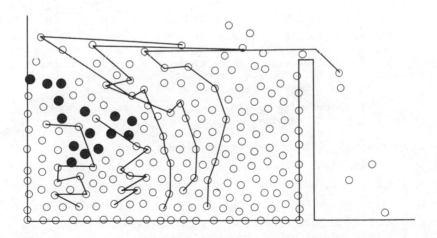

Fig. 5.29
t_{1600}

each drop was the same as each particle in the well, and if the initial velocity of each particle was taken to be $\vec{0}$, then the drop merely flowed to the right on top of the liquid in the well. The motion was entirely analogous to that of a wave rolling on top of a liquid base. Also, increasing the mass of each drop particle up to m = 10 led to motion which seemed to be physically excessive. Thus, though larger masses for drop particles can be introduced, the computer experiments indicated that a change to smaller Δt would probably yield results which were more accurate physically when m > 5.0.

5.7 POROUS FLOW

Porous flow study is of broad interest, for example, to geologists, ecologists, and civil engineers (see, e.g., [42], [79], and the numerous references contained therein). In general, it is the study of liquid transport through a solid or an earthlike conglomerate of small solids.

Let us consider a particular porous flow problem in which the liquid and the solid are as shown in Fig. 5.30. The liquid is contained within a square region above the triangular, porous ground section below it. The configuration is complicated by the presence of a solid shelf, shown below the right half of the liquid, through which the liquid cannot flow. The problem is to describe the flow of the liquid around the shelf and through the ground.

A total of 128 particles are chosen, 100 being liquid and 28 being solid. The parameter choices are H = 100, G = 0, q = 3, $\Delta t = 10^{-4}$, a local interaction distance of 0.25, and a velocity damping factor off the walls of 0.1. The mass of each liquid particle is taken to be unity. No motion is allowed for any solid particle, which enables one to enhance the liquid-solid interaction by taking the mass of each solid particle to be 0.2 (see [42] for the rationale of this choice). Gravity is fixed at $-980m_i$. The solid particles are arranged uniformly, as shown in Fig. 5.30 and all initial positions and velocities of the fluid particles are recorded in Table 5.8.

Figures 5.31-5.35 show the resulting flow from t_0 to t_{12000}. Initially, the fluid enters the porous area and quickly saturates the left corner, forming a small dead zone. Then, because the porous area has open horizontal channels, there is a rapid horizontal flow to the right, as is shown in Fig. 5.33, as the liquid follows a path of least resistance. The flow vertically then continues in the fashion shown in Fig. 5.34, and saturation is almost complete by time t_{12000}, as shown in Fig. 5.35. It is interesting to note, in this last figure, that the porous area just below the shelf still remains unoccupied.

Interesting variations of the above example were also run. If m_2 is the mass of each solid particle, then, setting m_2 = 0.05 or 0.1 resulted in relatively rapid porous flow, while setting m_2 = 0.5, 1.0 or 3.0 resulted in relatively slow flows. The choice m_2 = 10.0 resulted in no flow at all. Horizontal resetting of alternate rows of solid particles in Fig. 5.30, so that the porous area was characterized by vertical columns of particles and open vertical channels, resulted in rapid vertical flows rather than horizontal ones. In the examples with vertical open channels, the choice m_2 = 1 led to rapid vertical flow, but no horizontal flow. Thus, the left portion of the porous area became saturated quickly, but the solid particles were packed too closely in the vertical direction to allow penetration by the liquid particles, so that no flow to the right occurred. The case

TABLE 5.8 Ground Flow Initial Data

i	m_i	x	y	v_x	v_y
1	1.00000	-0.99990	0.00030	0.01750	0.03350
2	1.00000	-1.00000	0.18450	0.08590	-0.00140
3	1.00000	-0.99990	0.38670	0.12390	0.00780
4	1.00000	-0.99990	0.59440	0.07230	-0.02770
5	1.00000	-1.00000	0.81370	0.04830	-0.28770
6	1.00000	-0.99990	1.03250	0.12090	-0.00090
7	1.00000	-0.99990	1.28250	0.06940	0.00910
8	1.00000	-0.99990	1.50010	-0.85680	-0.22450
9	1.00000	-1.00000	1.75010	0.04370	0.02960
10	1.00000	-1.00000	2.00000	-0.11510	-0.05380
11	1.00000	-0.77540	0.0	0.02720	0.12990
12	1.00000	-0.52560	0.00010	-0.00770	0.13590
13	1.00000	-0.30610	0.00010	-0.00600	0.12310
14	1.00000	-0.09260	0.0	-0.12700	-0.58720
15	1.00000	1.00000	0.18620	-0.05830	-0.00200
16	1.00000	1.00000	0.37690	0.83630	-0.00440
17	1.00000	1.00000	0.59200	-0.11260	0.01240
18	1.00000	1.00000	0.81820	-0.04410	-0.04860
19	1.00000	1.00000	1.04650	-0.04670	-0.03930
20	1.00000	1.00000	1.27580	-0.04460	0.01240
21	1.00000	1.00000	1.50000	-0.08980	0.01910
22	1.00000	1.00000	1.75000	0.49830	-0.21750
23	1.00000	0.99990	2.00000	-0.01290	-0.00540
24	1.00000	-0.40250	0.21770	1.07210	-3.12640
25	1.00000	-0.62250	0.15430	-0.07290	-1.33340
26	1.00000	-0.78720	0.46290	-0.02100	0.85950
27	1.00000	-0.77600	0.92680	0.55330	1.27510
28	1.00000	-0.71670	1.35440	0.95080	-0.90450
29	1.00000	-0.81240	1.58470	0.47460	-0.38650
30	1.00000	-0.58270	1.56480	-4.22300	-1.48850
31	1.00000	-0.69800	1.78610	-0.96930	0.26940
32	1.00000	-0.79510	1.99990	0.00130	-0.05490
33	1.00000	-0.19980	0.22620	-0.62020	1.55790
34	1.00000	0.56850	0.35770	-0.60340	-0.05130
35	1.00000	-0.79380	0.24950	-0.29620	1.83360
36	1.00000	-0.78970	0.67860	-1.98360	-1.19850
37	1.00000	-0.81540	1.15940	0.84050	0.39330
38	1.00000	-0.35030	1.13210	-2.11790	0.90590
39	1.00000	-0.60090	1.13300	0.51920	0.39430
40	1.00000	-0.45100	1.77770	0.63110	0.18220
41	1.00000	-0.56610	2.00000	-0.01750	-0.09690
42	1.00000	-0.10450	0.42780	-1.45640	-0.24600
43	1.00000	-0.60820	0.78040	-0.28170	-1.45740
44	1.00000	-0.38520	0.70440	-1.80210	-1.66090
45	1.00000	-0.48260	0.93060	0.43770	0.00570
46	1.00000	-0.46670	1.34410	2.58530	0.62210
47	1.00000	-0.33370	1.55630	1.07310	-0.36560
48	1.00000	-0.21280	1.77520	0.09440	-1.78200
49	1.00000	0.13150	1.55290	-0.10810	-1.54960
50	1.00000	-0.33060	1.99860	-0.99100	-1.62130
51	1.00000	-0.34050	0.45990	-0.04530	-0.12260
52	1.00000	0.02440	0.21390	1.72610	-1.33930
53	1.00000	-0.18110	0.66480	1.60680	-0.79070

Arithmetic Applied Mathematics

TABLE 5.8 (Concluded)

i	m_i	x	y	v_x	v_y
54	1.00000	-0.24420	0.90680	0.97830	2.67730
55	1.00000	-0.21521	1.33420	0.12300	0.10320
56	1.00000	-0.09250	1.55130	-0.75730	-0.04570
57	1.00000	0.13760	1.06250	-3.60700	0.43600
58	1.00000	0.28210	1.35420	2.22790	0.51300
59	1.00000	-0.09990	1.99890	-0.35220	-1.12870
60	1.00000	0.39310	0.68460	-1.65790	1.77870
61	1.00000	0.38850	0.42500	3.45980	-0.22090
62	1.00000	0.13940	0.40540	-2.63570	0.72260
63	1.00000	-0.56880	0.56250	-0.68680	1.43560
64	1.00000	0.18170	0.81710	-0.05590	-2.67810
65	1.00000	0.03550	1.32210	-0.24930	0.39830
66	1.00000	0.34270	1.15810	0.00310	-2.08340
67	1.00000	0.01670	1.77720	0.41230	-0.98110
68	1.00000	0.13150	1.99980	-0.18170	1.52740
69	1.00000	0.26390	0.20860	-1.42660	0.47700
70	1.00000	0.50030	0.20090	-0.20200	2.26570
71	1.00000	-0.10130	1.11190	-0.56340	1.68090
72	1.00000	-0.03410	0.87060	1.78720	2.81500
73	1.00000	0.02760	0.62860	2.10450	-0.49440
74	1.00000	0.56890	1.06930	-1.08370	-1.42970
75	1.00000	0.59850	1.56000	-1.12270	0.36210
76	1.00000	0.24500	1.77570	1.30800	-0.03650
77	1.00000	0.35860	1.99840	-0.28720	1.42950
78	1.00000	0.75030	0.19830	-0.83960	2.59610
79	1.00000	0.77720	0.93260	-0.84330	0.18610
80	1.00000	0.23470	0.57660	-0.86370	1.39880
81	1.00000	0.78680	0.68720	0.04790	-1.16980
82	1.00000	0.36010	0.93370	-2.12340	-0.62710
83	1.00000	0.82010	1.60810	-0.79210	0.92540
84	1.00000	0.57800	0.82320	-1.28770	1.26140
85	1.00000	0.53280	1.31970	-0.00950	1.72490
86	1.00000	0.58820	2.00000	0.01290	-0.04740
87	1.00000	0.80040	0.44330	-1.66270	0.00190
88	1.00000	0.58820	0.57450	-0.68280	-0.26690
89	1.00000	0.77800	1.16120	0.27530	0.12400
90	1.00000	0.60860	0.36610	2.07950	-0.50330
91	1.00000	0.77450	1.38440	-1.32980	0.99790
92	1.00000	0.71380	1.78340	-0.81440	-0.06080
93	1.00000	0.48090	1.78000	-0.29750	1.06100
94	1.00000	0.37090	1.55910	-0.27390	0.65640
95	1.00000	0.80290	1.99990	-0.00020	-0.05660
96	1.00000	0.13600	0.0	-0.00360	0.11920
97	1.00000	0.36410	0.00010	0.00250	-1.15130
98	1.00000	0.59670	0.0	0.01270	0.08320
99	1.00000	0.80710	0.00010	-0.00050	0.10900
100	1.00000	0.99990	0.00010	-0.01380	0.01630

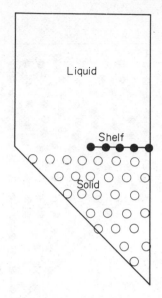

Fig. 5.30

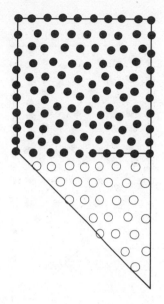

Fig. 5.31 - t_0

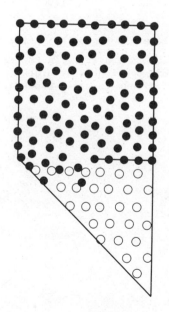

Fig. 5.32 - t_{200}

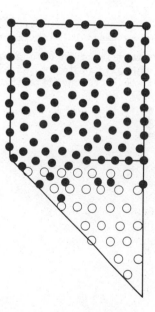

Fig. 5.33 - t_{800}

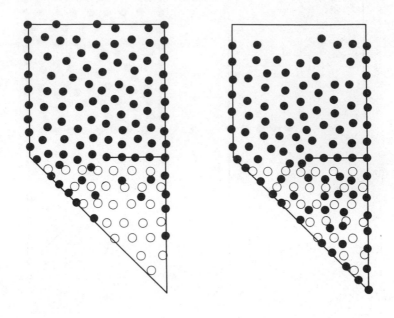

Fig. 5.34 - t_{2400} Fig. 5.35 - t_{12000}

of horizontal open channels with $m_2 = 1$ allows flow into the entire porous area because the primary force in this configuration is gravity. In the case of vertical channels, the primary force, after the left section is saturated, is liquid pressure, which never reaches a sufficient magnitude to allow penetration.

Chapter 6

Foundational Concepts of Special Relativity

6.1 INTRODUCTION

We turn now to the other major area of mechanics which is also deterministic, special relativity, and, as in the case of Newtonian mechanics, we will show how to formulate and develop the subject using only arithmetic. The special advantages of such a simplistic formulation will be discussed later. For the present, we will begin with a discussion of some important mathematical and physical preliminaries.

6.2 BASIC CONCEPTS

In making observations, it takes time for the light to travel from an observed event or object to the eye of the observer. In Newtonian mechanics, one assumes that such observations are made instantaneously. In special relativity we do account for the time it takes for light to travel. In so doing we will, for convenience, let c represent the speed of light in a vacuum [93].

In special relativity one uses rectangular, cartesian three-dimensional coordinate systems. The distance r between two points (x_1, y_1, z_1) and (x_2, y_2, z_2) in a given XYZ frame is determined by the classical Euclidean formula

$$r = [(x_2 - x_1)^2 + (y_2 - y_1)^2 + (z_2 - z_1)^2]^{\frac{1}{2}}. \qquad (6.1)$$

It is in general relativity where noncartesian coordinate systems are used and where nonrectangular distance formulas are fundamental. We will not consider any aspect of general relativity, which has less experimental support than does special relativity [7].

All reference frames in special relativity are assumed to be inertial, by which is meant the following ([7], [93]). In the absence of an external force, in an inertial system every particle which is initially at rest will remain at rest, while every particle which is initially in motion will continue that motion without change in speed or direction. It can be shown that every two inertial reference frames are related such that one is in relative uniform motion with respect to the other [88]. The basic assumption about inertial systems, called the axiom of relativity, and which is supported extensively by experimental evidence ([4], [7], [80],

[93]) is given as follows: All the laws of physics are the same in every inertial reference frame, and this includes the invariance of all physical constants, including the speed of light.

6.3 EVENTS AND A SPECIAL LORENTZ TRANSFORMATION

Consider now two rectangular, cartesian coordinate systems XYZ and X'Y'Z' which, intially, coincide and are, of course, inertial. Consider the X'Y'Z' system as being in uniform motion relative to the XYZ system. For the present, assume that this motion is in the X direction only and let the speed of X'Y'Z' relative to XYZ be u. We call XYZ the lab frame and X'Y'Z' the rocket frame.

For $\Delta t > 0$, let an observer in the lab frame make observations at the distinct times $t_k = k \, \Delta t$, $k = 0,1,2,\ldots$. Using an identical, synchronized clock, an observer in the rocket frame makes observations at the corresponding times t_k', where t_k' on the rocket clock corresponds to t_k on the lab clock.

Now, let particle P be at (x_k, y_k, z_k) at t_k in the lab frame and let it be at (x_k', y_k', z_k') in the rocket frame at the corresponding time t_k'. Then (x_k, y_k, z_k, t_k) and (x_k', y_k', z_k', t_k') are called <u>events</u> because each quadruple of points incorporates the knowledge of both position and time of observation. The precise mathematical formulas relating events were developed first by Lorentz [80] and later, in an alternate fashion, by Einstein [24]. These formulas are linear, are called the Lorentz transformation, and are given as follows:

$$x_k' = \frac{c(x_k - ut_k)}{(c^2 - u^2)^{\frac{1}{2}}}$$

$$y_k' = y_k$$

$$z_k' = z_k \qquad\qquad\qquad (6.2)$$

$$t_k' = \frac{c^2 t_k - ux_k}{c(c^2 - u^2)^{\frac{1}{2}}} \ .$$

Formulas (6.2) can, of course, be given in the equivalent form

$$x_k = \frac{c(x_k' + ut_k')}{(c^2 - u^2)^{\frac{1}{2}}}$$

$$y_k = y_k'$$

$$z_k = z_k' \qquad\qquad\qquad (6.3)$$

$$t_k = \frac{c^2 t_k' + ux_k'}{c(c^2 - u^2)^{\frac{1}{2}}} \ .$$

Observe, immediately, that the Lorentz transformation formulas involve only arithmetic operations, so that they themselves are consistent with the spirit in which
we will develop special relativity. Note also that it will be mathematically convenient in (6.2) to avoid division by zero, so that we will assume that

$$|u| < c. \tag{6.4}$$

Finally, observe that the classical results on the contraction of moving rods and
time dilation of moving clocks are a direct consequence of the Lorentz transformation. To deduce these, it is only important in discussing these apparently sur-
prising conclusions to be quite certain of the reference frames being used. Thus,
if one considers a rod in the <u>rocket</u> frame, whose end points have coordinates x_1'
and x_2' with $x_2' > x_1'$, then its length in the <u>rocket</u> frame is $x_2' - x_1'$.
Consider now an observer who is in the <u>lab</u> frame and who measures the length of the
rod, which is in motion relative to him, at the time $t_0 = 0$. Then, from (6.2),

$$x_2' - x_1' = \frac{c(x_2-x_1)}{(c^2-u^2)^{\frac{1}{2}}} , \tag{6.5}$$

which implies

$$x_2 - x_1 = \frac{(c^2-u^2)^{\frac{1}{2}}}{c} (x_2'-x_1') = (1-\frac{u^2}{c^2})^{\frac{1}{2}} (x_2'-x_1'). \tag{6.6}$$

But,

$$(1-\frac{u^2}{c^2})^{\frac{1}{2}} < 1,$$

so that

$$x_2 - x_1 < x_2' - x_1' \tag{6.7}$$

and the measured length (6.6), in the <u>lab</u> frame, is <u>less</u> than the measured length
in the <u>rocket</u> frame. As to time dilation, consider a clock which is at the origin
of the <u>rocket</u> frame and consider two consecutive time readings t_1' and t_2'.
Then, in the <u>rocket</u> frame, the time interval between the readings is $t_2' - t_1'$.
But, to an observer in the <u>lab</u> frame, the respective readings, from (6.3), are

$$t_1 = \frac{ct_1'}{(c^2-u^2)^{\frac{1}{2}}} , \qquad t_2 = \frac{ct_2'}{(c^2-u^2)^{\frac{1}{2}}} ,$$

and the interval between the readings is

$$t_2 - t_1 = \frac{c}{(c^2-u^2)^{\frac{1}{2}}} (t_2'-t_1') = \frac{1}{(1-\frac{u^2}{c^2})^{\frac{1}{2}}} (t_2'-t_1') . \tag{6.8}$$

But,

$$\frac{1}{(1-\frac{u^2}{c^2})^{\frac{1}{2}}} > 1,$$

so that

$$t_2 - t_1 > t_2' - t_1',$$
(6.9)

and the measured time interval (6.8), in the lab frame, is greater than the measured time interval in the rocket frame.

6.4   A GENERAL LORENTZ TRANSFORMATION

From the dynamical point of view, that is, when one is concerned with the motion of a particle, we will require a more general motion of the rocket frame relative to the lab frame than that described in Section 6.3. The resulting Lorentz transformation formulas will, of course, be more extensive than those given by (6.2). In this section we will summarize the most important such generalization ([4], [67]).

Under the same assumptions as those of Section 6.3, let us now suppose that the relative uniform velocity of the rocket frame X'Y'Z' relative to the lab frame XYZ is given by $\vec{u} = (u_1, u_2, u_3)$.

Let

$$\vec{\beta} = (\beta_1, \beta_2, \beta_3) = \vec{u}/c$$
(6.10)

$$u^2 = u_1^2 + u_2^2 + u_3^2 = c^2(\beta_1^2 + \beta_2^2 + \beta_3^2) = c^2\beta^2$$
(6.11)

$$\gamma = (1-\beta^2)^{-\frac{1}{2}}.$$
(6.12)

With regard to events (x_k, y_k, z_k, t_k) and (x_k', y_k', z_k', t_k'), let

$$\vec{r}_k = \begin{pmatrix} x_k \\ y_k \\ z_k \\ t_k \end{pmatrix} \qquad , \qquad \vec{r}_k' = \begin{pmatrix} x_k' \\ y_k' \\ z_k' \\ t_k' \end{pmatrix}.$$
(6.13)

Then the Lorentz transformation $L^* = (L_{ij}^*)$, i = 1,2,3,4; j = 1,2,3,4, is given by [4, p. 74]:

$$\vec{r}_k' = L^*\vec{r}_k,$$
(6.14)

where

$$(L^*_{ij}) = \begin{pmatrix} 1+\beta_1^2\dfrac{\gamma^2}{\gamma+1} & \beta_1\beta_2\dfrac{\gamma^2}{\gamma+1} & \beta_1\beta_3\dfrac{\gamma^2}{\gamma+1} & -c\beta_1\gamma \\[2ex] \beta_1\beta_2\dfrac{\gamma^2}{\gamma+1} & 1+\beta_2^2\dfrac{\gamma^2}{\gamma+1} & \beta_2\beta_3\dfrac{\gamma^2}{\gamma+1} & -c\beta_2\gamma \\[2ex] \beta_1\beta_3\dfrac{\gamma^2}{\gamma+1} & \beta_2\beta_3\dfrac{\gamma^2}{\gamma+1} & 1+\beta_3^2\dfrac{\gamma^2}{\gamma+1} & -c\beta_3\gamma \\[2ex] -\dfrac{\beta_1}{c}\gamma & -\dfrac{\beta_2}{c}\gamma & -\dfrac{\beta_3}{c}\gamma & \gamma \end{pmatrix} . \qquad (6.15)$$

The transformation (6.14) is convenient from the physical point of view. From the geometric point of view, as will be indicated later, a more convenient form can be given as follows. Let new coordinates, called Minkowski coordinates, defined by

$$\left.\begin{array}{llll} x_{1,k} = x_k, & x_{2,k} = y_k, & x_{3,k} = z_k, & x_{4,k} = ict_k \\[1ex] x_{1,k}' = x_k', & x_{2,k}' = y_k', & x_{3,k}' = z_k', & x_{4,k}' = ict_k' \end{array}\right\} . \qquad (6.16)$$

Let

$$\vec{R}_k = \begin{pmatrix} x_{1,k} \\ x_{2,k} \\ x_{3,k} \\ x_{4,k} \end{pmatrix}, \qquad\qquad \vec{R}_k' = \begin{pmatrix} x_{1,k}' \\ x_{2,k}' \\ x_{3,k}' \\ x_{4,k}' \end{pmatrix} . \qquad (6.17)$$

Then the Lorentz transformation $L = (L_{ij})$ is given by [4, p. 74]:

$$\vec{R}_k' = L\vec{R}_k, \qquad (6.18)$$

where

$$(L_{ij}) = \begin{pmatrix} 1+\beta_1^2\dfrac{\gamma^2}{\gamma+1} & \beta_1\beta_2\dfrac{\gamma^2}{\gamma+1} & \beta_1\beta_3\dfrac{\gamma^2}{\gamma+1} & i\beta_1\gamma \\[2ex] \beta_1\beta_2\dfrac{\gamma^2}{\gamma+1} & 1+\beta_2^2\dfrac{\gamma^2}{\gamma+1} & \beta_2\beta_3\dfrac{\gamma^2}{\gamma+1} & i\beta_2\gamma \\[2ex] \beta_1\beta_3\dfrac{\gamma^2}{\gamma+1} & \beta_2\beta_3\dfrac{\gamma^2}{\gamma+1} & 1+\beta_3^2\dfrac{\gamma^2}{\gamma+1} & i\beta_3\gamma \\[2ex] -i\beta_1\gamma & -i\beta_2\gamma & -i\beta_3\gamma & \gamma \end{pmatrix} . \qquad (6.19)$$

With regard to (6.19), note that

$$\sum_{j=1}^{4} L_{ij}L_{kj} = \delta_{i,k} \tag{6.20}$$

$$\sum_{j=1}^{4} L_{ji}L_{jk} = \delta_{i,k} , \tag{6.21}$$

where $\delta_{i,k}$ is the Kronecker δ, and that

$$L^{T}L = LL^{T} = I , \tag{6.22}$$

where L^{T} is the transpose of L and I is the identity.

Even more general transformations can be developed by rotating axes, and so forth [67], but these are of little practical importance and hence will not be given here.

Chapter 7

Arithmetic Special Relativistic Mechanics in One Space Dimension

7.1 INTRODUCTION

For simplicity, let us begin our study of arithmetic special relativity by assuming that the rocket frame is in uniform motion relative to the lab frame in the X direction only, as described in Section 6.3. For completeness, recall that events (x_k, y_k, z_k, t_k) and (x_k', y_k', z_k', t_k') are then related by the Lorentz transformation:

$$x_k' = \frac{c(x_k - ut_k)}{(c^2 - u^2)^{\frac{1}{2}}}$$

$$y_k' = y_k$$

$$z_k' = z_k \qquad\qquad (7.1)$$

$$t_k' = \frac{c^2 t_k - ux_k}{c(c^2 - u^2)^{\frac{1}{2}}}\ ,$$

or, equivalently, by

$$x_k = \frac{c(x_k' + ut_k')}{(c^2 - u^2)^{\frac{1}{2}}}$$

$$y_k = y_k'$$

$$z_k = z_k' \qquad\qquad (7.1')$$

$$t_k = \frac{c^2 t_k' + ux_k'}{c(c^2 - u^2)^{\frac{1}{2}}}\ .$$

115

For later convenience, note also that throughout the remainder of this chapter, the symbol Δ will represent a forward difference operator defined by

$$\Delta F(k) = F(k+1) - F(k). \tag{7.2}$$

7.2 PROPER TIME

It is common in relativity to call the reference frame in which an object is at rest the _proper_ _frame_ and the time of an event in this frame, the _proper_ _time_. Formula (6.8) relates the proper time $(t_2'-t_1')$ in the rocket frame to the "improper" time (t_2-t_1) in the lab frame, since the clock was assumed to have zero velocity relative to the rocket frame. More formally, the concept of proper time can be developed in the following simple way.

In the rocket frame, the proper time τ_k of event (x_k',y_k',z_k',t_k') is defined by

$$\tau_k = (c^2 {t_k'}^2 - {x_k'}^2 - {y_k'}^2 - {z_k'}^2)^{\frac{1}{2}} \tag{7.3}$$

provided that

$$c^2 {t_k'}^2 - {x_k'}^2 - {y_k'}^2 - {z_k'}^2 > 0. \tag{7.4}$$

Remarkably enough, τ_k is invariant under the Lorentz transformation, that is, τ_k is also the proper time of event (x_k,y_k,z_k,t_k);

$$\tau_k = (c^2 {t_k}^2 - {x_k}^2 - {y_k}^2 - {z_k}^2)^{\frac{1}{2}}. \tag{7.5}$$

To prove this, note that from (7.1)

$$c^2 {t_k'}^2 - {x_k'}^2 - {y_k'}^2 - {z_k'}^2 = c^2 \left(\frac{c^2 t_k - u x_k}{c(c^2-u^2)^{\frac{1}{2}}} \right)^2 - \left(\frac{c(x_k - u t_k)}{(c^2-u^2)^{\frac{1}{2}}} \right)^2 - {y_k}^2 - {z_k}^2$$

$$= \frac{(c^2-u^2)(c^2 {t_k}^2 - {x_k}^2)}{(c^2-u^2)} - {y_k}^2 - {z_k}^2$$

$$= c^2 {t_k}^2 - {x_k}^2 - {y_k}^2 - {z_k}^2 ,$$

from which the invariance follows immediately. Note, also, that this invariance and (7.4) imply

$$c^2 {t_k}^2 - {x_k}^2 - {y_k}^2 - {z_k}^2 > 0. \tag{7.6}$$

The definition of proper time yields a certain unity of space and time, often called _space-time_ [93] because the units of ct_k in (7.5), like the units of x_k, y_k and z_k, are units of length.

Note also that, as above, the quantity $\delta\tau_k$ defined by

$$\delta\tau_k = [c^2(\Delta t_k')^2 - (\Delta x_k')^2 - (\Delta y_k')^2 - (\Delta z_k')^2]^{\frac{1}{2}} \qquad (7.7)$$

is also invariant, that is, under the Lorentz transformation,

$$\delta\tau_k = [c^2(\Delta t_k)^2 - (\Delta x_k)^2 - (\Delta y_k)^2 - (\Delta z_k)^2]^{\frac{1}{2}}. \qquad (7.8)$$

As in (7.4), we will assume that

$$c^2(\Delta t_k')^2 - (\Delta x_k')^2 - (\Delta y_k')^2 - (\Delta z_k')^2 > 0, \qquad (7.9)$$

which is equivalent to

$$c^2(\Delta t_k)^2 - (\Delta x_k)^2 - (\Delta y_k)^2 - (\Delta z_k)^2 > 0. \qquad (7.10)$$

The quantity $\delta\tau_k$ is then called the proper time between successive events (x_k',y_k',z_k',t_k') and $(x_{k+1}',y_{k+1}',z_{k+1}',t_{k+1}')$. Of course, it is also the proper time between successive events (x_k,y_k,z_k,t_k) and $(x_{k+1},y_{k+1},z_{k+1},t_{k+1})$.

Note that $\delta\tau_k$ is not, in general, the same as $\Delta\tau_k$.

7.3 VELOCITY AND ACCELERATION

We wish now to begin to study the motion of a particle P and we assume, in this chapter, that its motion is also in the X direction only. To analyze the motion of P, it will be convenient, in analogy with our development of Newtonian mechanics, to have arithmetic concepts of velocity, of acceleration and of a dynamical difference equation which is symmetric (invariant) under the Lorentz transformation. Let us then develop the concepts of velocity and acceleration first.

Recall that we are assuming that P is at (x_k,y_k,z_k) at t_k in the lab frame and at (x_k',y_k',z_k') at the corresponding time t_k' in the rocket frame. Then, at t_k, P's velocity $v(t_k) = v_k$ and acceleration $a(t_k) = a_k$ are defined in the lab frame by

$$v_k = \frac{\Delta x_k}{\Delta t_k} \qquad (7.11)$$

$$a_k = \frac{\Delta v_k}{\Delta t_k} \qquad (7.12)$$

By the axiom of relativity, in the rocket frame one must define v_k' and a_k' at t_k' by

$$v_k' = \frac{\Delta x_k'}{\Delta t_k'} \qquad (7.11')$$

$$a_k' = \frac{\Delta v_k'}{\Delta t_k'} \; . \tag{7.12'}$$

In order to develop connecting relationships between v_k and v_k' and between a_k and a_k', note first that (7.1) and (7.2) imply

$$\Delta x_k' = \frac{c(\Delta x_k - u\Delta t_k)}{(c^2 - u^2)^{\frac{1}{2}}} \tag{7.13}$$

$$\Delta y_k' = \Delta y_k \tag{7.14}$$

$$\Delta z_k' = \Delta z_k \tag{7.15}$$

$$\Delta t_k' = \frac{(c^2 \Delta t_k - u\Delta x_k)}{c(c^2 - u^2)^{\frac{1}{2}}} \; . \tag{7.16}$$

Hence, (7.11'), (7.13) and (7.16) imply

$$v_k' = \frac{c(\Delta x_k - u\Delta t_k)}{(c^2 - u^2)^{\frac{1}{2}}} \; \frac{c(c^2 - u^2)^{\frac{1}{2}}}{c^2 \Delta t_k - u\Delta x_k} = \frac{c^2(\Delta x_k - u\Delta t_k)}{c^2 \Delta t_k - u\Delta x_k} \; ,$$

so that

$$v_k' = \frac{c^2(v_k - u)}{c^2 - uv_k} \; , \tag{7.17}$$

while, (7.12'), (7.16) and (7.17) imply

$$a_k' = \left[\frac{c^2(v_{k+1} - u)}{c^2 - uv_{k+1}} - \frac{c^2(v_k - u)}{c^2 - uv_k} \right] \cdot \frac{c(c^2 - u^2)^{\frac{1}{2}}}{c^2 \Delta t_k - u\Delta x_k}$$

$$= \frac{c^2(c^2 v_{k+1} - c^2 v_k - u^2 v_{k+1} + u^2 v_k)}{(c^2 - uv_{k+1})(c^2 - uv_k)} \cdot \frac{c(c^2 - u^2)^{\frac{1}{2}}}{c^2 \Delta t_k - u\Delta x_k}$$

$$= \frac{c^2(v_{k+1} - v_k)(c^2 - u^2)}{(c^2 - uv_{k+1})(c^2 - uv_k)} \cdot \frac{c(c^2 - u^2)^{\frac{1}{2}}}{c^2 \Delta t_k - u\Delta x_k}$$

$$= \frac{c^3(c^2 - u^2)^{3/2}}{(c^2 - uv_{k+1})(c^2 - uv_k)^2} \cdot \frac{v_{k+1} - v_k}{\Delta t_k} \; ,$$

so that

$$a_k' = \frac{c^3(c^2-u^2)^{3/2}}{(c^2-uv_{k+1})(c^2-uv_k)^2}\, a_k \,. \tag{7.18}$$

Of course, (7.17) is equivalent to

$$v_k = \frac{c^2(v_k'+u)}{c^2+uv_k'} \,, \tag{7.17'}$$

while (7.18) is equivalent to

$$a_k = \frac{c^3(c^2-u^2)^{3/2}}{(c^2+uv_{k+1}')(c^2+uv_k')^2}\, a_k' \,. \tag{7.18'}$$

Note also that for the present type of restricted motions, (7.9), (7.10), (7.11) and (7.11') imply

$$|v_k| < c \tag{7.19}$$

$$|v_k'| < c. \tag{7.19'}$$

In preparation for the study of a dynamical difference equation, let us briefly examine the concepts of rest mass and linear momentum.

7.4   REST MASS AND MOMENTUM

From experiments in which electrons are accelerated until their velocities are close to the speed of light, we know that the masses of such particles increase with velocity [26]. For this reason, let $m(k)$ be the mass of particle P which has velocity v_k at time t_k in the lab frame. Also, let $m'(k)$ be the mass of P when its velocity is v_k' at corresponding time t_k' in the rocket frame. Then the linear momentum p_k of P at t_k in the lab frame is defined by

$$p_k = m(k)v_k. \tag{7.20}$$

By the axiom of relativity, P's linear momentum in the rocket frame must be defined by

$$p_k' = m'(k)v_k'. \tag{7.20'}$$

It is now worth noting that formula (7.17) for the transformation of velocities is identical with that of continuous relativistic mechanics ([80], [93]). Thus, the usual arguments [93], in which identical objects undergo elastic collisions again yield the results that, if momentum is to be conserved, then it is necessary that $m(k)$ and $m'(k)$ satisfy

$$m(k) = \frac{cm_0}{(c^2-v_k^2)^{\frac{1}{2}}} \tag{7.21}$$

$$m'(k) = \frac{cm_0}{(c^2 - v_k'^2)^{\frac{1}{2}}} , \qquad\qquad (7.21')$$

where m_0 is called the <u>rest</u> <u>mass</u> of P. The rest mass is also called the proper mass because it is P's mass in a system in which P is at rest. We will continue to assume the validity of (7.21) and (7.21') and we will assume that the law of conservation of momentum is valid, whether or not the colliding objects are identical and whether or not the collision is elastic.

7.5 THE DYNAMICAL DIFFERENCE EQUATION

The Newtonian dynamical equation $F_k = ma_k$ is <u>not</u> symmetric with respect to the Lorentz transformation. Thus, to preserve the axiom of relativity, we must define a new dynamical equation. This equation, together with (7.11) and (7.12), or (7.11') and (7.12'), will enable one to determine the motion of a given particle from given initial data, i.e., from its position and velocity at t_0.

In the rocket frame, the equation which we elect to use is [40]:

$$F_k' = \frac{c^2 m'(k)}{[(c^2 - v_k'^2)(c^2 - v_{k+1}'^2)]^{\frac{1}{2}}} \cdot \frac{\Delta v_k'}{\Delta t_k'} . \qquad\qquad (7.22)$$

Let us first establish symmetry, that is, that under the Lorentz transformation, $F_k' = F_k$ implies

$$F_k = \frac{c^2 m(k)}{[(c^2 - v_k^2)(c^2 - v_{k+1}^2)]^{\frac{1}{2}}} \cdot \frac{\Delta v_k}{\Delta t_k} . \qquad\qquad (7.23)$$

To do this, note that (7.16), (7.17), (7.18), (7.21), (7.21') and (7.22) imply

$$F_k' = \frac{c^3 m_0}{(c^2 - v_k'^2)(c^2 - v_{k+1}'^2)^{\frac{1}{2}}} \cdot \frac{v_{k+1}' - v_k'}{\Delta t_k'}$$

$$= \frac{c^3 m_0}{\left[c^2 - \dfrac{c^4(v_k - u)^2}{(c^2 - uv_k)^2}\right]\left[c^2 - \dfrac{c^4(v_{k+1} - u)^2}{(c^2 - uv_{k+1})}\right]^{\frac{1}{2}}} \cdot \frac{\dfrac{c^2(v_{k+1} - u)}{c^2 - uv_{k+1}} - \dfrac{c^2(v_k - u)}{c^2 - uv_k}}{\dfrac{c^2 \Delta t_k - u \Delta x_k}{c(c^2 - u^2)^{\frac{1}{2}}}}$$

$$= \frac{c^3 m_0 [(v_{k+1} - u)(c^2 - uv_k) - (v_k - u)(c^2 - uv_{k+1})](c^2 - u^2)^{\frac{1}{2}}}{[(c^2 - uv_k)^2 - c^2(v_k - u)^2][(c^2 - uv_{k+1})^2 - c^2(v_{k+1} - u)^2]^{\frac{1}{2}} \Delta t_k}$$

$$= \frac{c^3 m_0 (v_{k+1} - v_k)(c^2 - u^2)^{3/2}}{(c^2 - u^2)(c^2 - v_k^2)[(c^2 - u^2)(c^2 - v_{k+1}^2)]^{\frac{1}{2}} \Delta t_k}$$

$$= \frac{c^3 m_0}{(c^2 - v_k^2)(c^2 - v_{k+1}^2)^{\frac{1}{2}}} \cdot \frac{v_{k+1} - v_k}{\Delta t_k}$$

$$= \frac{c^2 m(k)}{[(c^2 - v_k^2)(c^2 - v_{k+1}^2)]^{\frac{1}{2}}} \cdot \frac{\Delta v_k}{\Delta t_k} \, ,$$

from which (7.23) follows.

Note that taking limits in, say, (7.23), yields the particular form

$$F = \frac{c^2 m}{c^2 - v^2} \frac{dv}{dt}$$

of the classical and invariant relativistic differential equation

$$F = \frac{d}{dt}(mv), \tag{7.24}$$

where m is defined by (7.21), but with the index k deleted [26].

7.6 ENERGY

At time t_k in the lab frame, the total energy E of particle P is defined by

$$E = m(k)c^2. \tag{7.25}$$

Heuristic motivation for this definition follows from experiments in which matter is annihilated, that is, converted totally to energy [26]. Thus, when an electron and a positron come together at rest, each with rest mass m_0, they disintegrate and the two emerging gamma rays each has $m_0 c^2$ measured energy.

From (7.21) and (7.25), then,

$$E = \frac{c^2 m_0}{\left(1 - \frac{v_k^2}{c^2}\right)^{\frac{1}{2}}} \cdot \tag{7.26}$$

By (7.19) and (7.26),

$$E = c^2 m_0 \left(1 + \frac{1}{2} \frac{v_k^2}{c^2} + \dots \right),$$

122   Arithmetic Applied Mathematics

or,

$$E = c^2 m_0 + \frac{1}{2} m_0 v_k^2 + \dots . \qquad (7.27)$$

The quantity $\frac{1}{2} m_0 v_k^2$ is, of course, the classical Newtonian kinetic energy. For the special case $v_k = 0$, (7.27) reduces to

$$E_0 = m_0 c^2, \qquad (7.28)$$

which is called the _rest energy_, or proper energy, of P.

Another convenient formula for expressing energy is

$$E = m_0 c^3 \frac{\Delta t_k}{\delta \tau_k} , \qquad (7.29)$$

since $y_k = z_k = 0$ under present assumptions and

$$m_0 c^3 \frac{\Delta t_k}{\delta \tau_k} = \frac{m_0 c^3 \Delta t_k}{(c^2 \Delta t_k^2 - \Delta x_k^2 - \Delta y_k^2 - \Delta z_k^2)^{\frac{1}{2}}} = \frac{m_0 c^2}{(1 - \frac{v_k^2}{c^2})^{\frac{1}{2}}} .$$

Finally, let us derive relationships which connect energy and momentum. Elimination of $m(k)$ between (7.20) and (7.25) yields

$$p_k c^2 = v_k E. \qquad (7.30)$$

A second interesting relationship is

$$E^2 = p_k^2 c^2 + m_0^2 c^4, \qquad (7.31)$$

which follows from (7.20), (7.21), and (7.25), since

$$p_k^2 c^2 + m_0^2 c^4 = [m(k) v_k]^2 c^2 + m_0^2 c^4$$

$$= \frac{m_0^2 c^4 v_k^2}{c^2 - v_k^2} + m_0^2 c^4$$

$$= \frac{m_0^2 c^6}{c^2 - v_k^2}$$

$$= [m(k)]^2 c^4 .$$

The special significance of (7.31) is that it and the conservation of linear momentum imply the conservation of energy, which is why no special attention is directed toward the question of energy conservation. In relativistic mechanics, energy conservation is a direct consequence of momentum conservation.

7.7 THE MOMENTUM-ENERGY VECTOR

For all practical purposes, the restricted type motion studied in this chapter never requires consideration of y_k and z_k. For this reason, we will restrict attention in this section to the event (x_k, t_k), rather than to (x_k, y_k, z_k, t_k). The event (x_k, t_k) maps into the event (x_k', t_k') under the Lorentz transformation.

Also, thus far we have not placed any emphasis on any particular set of measurement units. In this connection, we will now be specific in the following way. Let

$$E^* = \frac{E}{c^2} \tag{7.32}$$

be a normalized energy in the sense that the units of E^*, by (7.25), are units of mass. Attention will be directed to E^*, rather than to E.

Our present purpose is to show that the number couple (p_k, E^*), where p_k is given by (7.20) and E^* is given by (7.32) is a vector, called the momentum-energy vector. Precisely, this means that (p_k, E^*) maps under the Lorentz transformation exactly like (x_k, t_k). Thus, from (7.1), we wish to show that

$$p_k' = \frac{c(p_k - uE^*)}{(c^2 - u^2)^{\frac{1}{2}}} \tag{7.33}$$

$$E^{*\prime} = \frac{c^2 E^* - u p_k}{c(c^2 - u^2)^{\frac{1}{2}}} . \tag{7.34}$$

From (7.17), (7.20)-(7.21') and (7.32)

$$p_k' = m'(k) v_k'$$

$$= \frac{c m_0}{(c^2 - v_k^2)^{\frac{1}{2}}} \cdot \frac{c^2 - u v_k}{c(c^2 - u^2)^{\frac{1}{2}}} \cdot \frac{c^2(v_k - u)}{c^2 - u v_k}$$

$$= \frac{c m(k)(v_k - u)}{(c^2 - u^2)^{\frac{1}{2}}}$$

$$= \frac{c[m(k) v_k - E^* u]}{(c^2 - u^2)^{\frac{1}{2}}} ,$$

from which (7.33) follows immediately. Then, from (7.17), (7.20), (7.21'), (7.25) and (7.32),

$$E*' = m'(k)$$

$$= \frac{cm_0}{(c^2-v_k'^2)^{\frac{1}{2}}}$$

$$= \frac{cm_0(c^2-uv_k)}{c(c^2-u^2)^{\frac{1}{2}}(c^2-v_k^2)^{\frac{1}{2}}}$$

$$= \frac{m(k)(c^2-uv_k)}{c(c^2-u^2)^{\frac{1}{2}}}$$

$$= \frac{c^2E*-up_k}{c(c^2-u^2)^{\frac{1}{2}}}$$

which establishes (7.34).

7.8 REMARKS

We have shown in Sections 7.1–7.7 how to formulate the basic physical concepts of relativity using only arithmetic processes. In particular, differences and difference quotients played a major role. Attention was restricted, for simplicity, to a very special class of particle and rocket frame motions, but, even so, all the basic consequences related to momentum, symmetry, energy, and momentum-energy vectors were deduced within this arithmetic framework. We will next extend the results of this chapter to more general motions and then show how initial value problems can be solved in a Lorentz invariant fashion using modern digital computers.

Chapter 8

Arithmetic Special Relativistic Mechanics in Three Space Dimensions

8.1 INTRODUCTION

The superiority of relativistic mechanics over Newtonian mechanics seems to be accepted almost universally, and for several very good reasons. The existence of rest energy is not a consequence of Newtonian mechanics and the application and control of this resource has been, and continues to be, of vital interest and importance. Also, from the electromagnetic wave point of view, it was in Einstein's fundamental paper of 1905 that Maxwell's equations were shown to be Lorentz invariant. Nevertheless, from the purely particle point of view, serious shortcomings are encountered in the relativistic approach. For example, preclusion of simultaneity prevents the formulation of any N-body problem, thereby further precluding modeling of the solar system and of galaxies. Further, the fundamental relativistic dynamical equation (7.24) is not Lorentz invariant in more than one space dimension, and one is thereby forced into a spacewise extension of relativity which is more geometric than physical. It must be with such limitations in mind that we now extend to three space dimensions the ideas and results established in Chapter 7.

8.2 VELOCITY, ACCELERATION, AND PROPER TIME

Consider now two rectangular, cartesian coordinate systems XYZ and X'Y'Z' which, initially, coincide. Let X'Y'Z', the rocket frame, be in relative uniform motion with respect to XYZ, the lab frame, and let this relative velocity be $\vec{u} = (u_1, u_2, u_3)$, as in Section 6.4.

Assume that particle P is in motion in the lab frame and at time t_k is at (x_k, y_k, z_k). Then P's velocity v_k and acceleration a_k at time t_k are defined by

$$\vec{v}_k = \begin{pmatrix} v_{1,k} \\ v_{2,k} \\ v_{3,k} \end{pmatrix} = \begin{pmatrix} \dfrac{\Delta x_k}{\Delta t_k} \\ \dfrac{\Delta y_k}{\Delta t_k} \\ \dfrac{\Delta z_k}{\Delta t_k} \end{pmatrix}, \tag{8.1}$$

$$\vec{a}_k = \begin{pmatrix} a_{1,k} \\ a_{2,k} \\ a_{3,k} \end{pmatrix} = \begin{pmatrix} \dfrac{\Delta v_{1,k}}{\Delta t_k} \\ \dfrac{\Delta v_{2,k}}{\Delta t_k} \\ \dfrac{\Delta v_{3,k}}{\Delta t_k} \end{pmatrix}. \tag{8.2}$$

By the axiom of relativity, P's velocity $\vec{v}_k{}'$ and acceleration $\vec{a}_k{}'$ in the rocket frame at corresponding time $t_k{}'$ are defined by

$$\vec{v}_k{}' = \begin{pmatrix} v_{1,k}{}' \\ v_{2,k}{}' \\ v_{3,k}{}' \end{pmatrix} = \begin{pmatrix} \dfrac{\Delta x_k{}'}{\Delta t_k{}'} \\ \dfrac{\Delta y_k{}'}{\Delta t_k{}'} \\ \dfrac{\Delta z_k{}'}{\Delta t_k{}'} \end{pmatrix}, \tag{8.1'}$$

$$\vec{a}_k{}' = \begin{pmatrix} a_{1,k}{}' \\ a_{2,k}{}' \\ a_{3,k}{}' \end{pmatrix} = \begin{pmatrix} \dfrac{\Delta v_{1,k}{}'}{\Delta t_k{}'} \\ \dfrac{\Delta v_{2,k}{}'}{\Delta t_k{}'} \\ \dfrac{\Delta v_{3,k}{}'}{\Delta t_k{}'} \end{pmatrix}. \tag{8.2'}$$

The respective magnitudes v_k, $v_k{}'$, a_k, $a_k{}'$ of $\vec{v}_k$, $\vec{v}_k{}'$, $\vec{a}_k$ and $\vec{a}_k{}'$ are defined in the customary way by

$$v_k{}^2 = v_{1,k}{}^2 + v_{2,k}{}^2 + v_{3,k}{}^2,$$

$$v_k{}'^2 = v_{k,1}{}'^2 + v_{k,2}{}'^2 + v_{k,3}{}'^2,$$

$$a_k{}^2 = a_{1,k}{}^2 + a_{2,k}{}^2 + a_{3,k}{}^2,$$

$$a_k{}'^2 = a_{1,k}{}'^2 + a_{2,k}{}'^2 + a_{3,k}{}'^2.$$

The quantity τ_k, defined in the lab frame by

$$\tau_k = (c^2 t_k{}^2 - x_k{}^2 - y_k{}^2 - z_k{}^2)^{\frac{1}{2}} \tag{8.3}$$

is invariant under L*, given by (6.15), since

$$c^2 t_k'^2 - x_k'^2 - y_k'^2 - z_k'^2 = c^2 t_k^2 - x_k^2 - y_k^2 - z_k^2 .$$

When

$$c^2 t_k^2 - x_k^2 - y_k^2 - z_k^2 > 0, \qquad (8.4)$$

τ_k is called the proper time of event (x_k, y_k, z_k, t_k), and, throughout, we assume that (8.4) is valid for all k. The quantity $\delta \tau_k$, defined by

$$\delta \tau_k = [c^2 (\Delta t_k)^2 - (\Delta x_k)^2 - (\Delta y_k)^2 - (\Delta z_k)^2]^{\frac{1}{2}} \qquad (8.5)$$

is, similarly, an invariant of L* and is called the proper time between successive events (x_k, y_k, z_k, t_k) and $(x_{k+1}, y_{k+1}, z_{k+1}, t_{k+1})$. Throughout, we assume, of course, that in (8.5)

$$c^2 (\Delta t_k)^2 - (\Delta x_k)^2 - (\Delta y_k)^2 - (\Delta z_k)^2 > 0. \qquad (8.6)$$

Note that (8.6) implies that

$$c^2 - \frac{(\Delta x_k)^2}{(\Delta t_k)^2} - \frac{(\Delta y_k)^2}{(\Delta t_k)^2} - \frac{(\Delta z_k)^2}{(\Delta t_k)^2} = c^2 - v_k^2 > 0, \qquad (8.7)$$

so that

$$|v_k| < c. \qquad (8.8)$$

For later convenience, note also that (8.8) and the invariance of $\delta \tau_k$ yield

$$\delta \tau_k = \Delta t_k [c^2 - v_k^2]^{\frac{1}{2}} = \Delta t_k' [c^2 - v_k'^2]^{\frac{1}{2}}. \qquad (8.9)$$

Finally, note that

$$v_{j,k}' = \frac{L^*_{j1} v_{1,k} + L^*_{j2} v_{2,k} + L^*_{j3} v_{3,k} + L^*_{j4}}{L^*_{41} v_{1,k} + L^*_{42} v_{2,k} + L^*_{43} v_{3,k} + L^*_{44}}, \quad j = 1,2,3, \qquad (8.10)$$

which generalizes (7.17) and shows again that the usual concept of velocity is not a vector with respect to the Lorentz transformation. It is the convenience derived when velocity is a vector with respect to the Lorentz transformation which motivates the abandonment of the physically meaningful definitions (8.1) and (8.2) and the introduction of new velocity and acceleration concepts in terms of Minkowski coordinates.

8.3 MINKOWSKI SPACE

Recall that Minkowski coordinates are defined in (6.16) by

$$\left.\begin{array}{llll} x_{1,k} = x_k, & x_{2,k} = y_k, & x_{3,k} = z_k, & x_{4,k} = ict_k \\[2mm] x_{1,k}' = x_k', & x_{2,k}' = y_k', & x_{3,k}' = z_k', & x_{4,k}' = ict_k' \end{array}\right\}, \qquad (8.11)$$

and observe that in Minkowski coordinates

$$\tau_k = [-\sum_{i=1}^{4} (x_{i,k})^2]^{\frac{1}{2}} = [-\sum_{i=1}^{4} (x_{i,k}')^2]^{\frac{1}{2}}, \qquad (8.12)$$

$$\delta\tau_k = [-\sum_{i=1}^{4} (\Delta x_{i,k})^2]^{\frac{1}{2}} = [-\sum_{i=1}^{4} (\Delta x_{i,k}')^2]^{\frac{1}{2}}. \qquad (8.13)$$

One now converts the laboratory frame, for example, into a Minkowski space as follows. Minkowski space is the set of quadruples $(x_{1,k}, x_{2,k}, x_{3,k}, x_{4,k})$ with the distance d between any two such quadruples, say, $(x_{1,k}, x_{2,k}, x_{3,k}, x_{4,k})$ and $(X_{1,k}, X_{2,k}, X_{3,k}, X_{4,k})$ being defined by

$$d = [-\sum_{i=1}^{4} (x_{i,k} - X_{i,k})^2]^{\frac{1}{2}}. \qquad (8.14)$$

In this geometric space, τ_k is the distance between $(x_{1,k}, x_{2,k}, x_{3,k}, x_{4,k})$ and the origin $(0,0,0,0)$.

We assume completely similar definitions in the rocket frame so that it too can be considered to be a Minkowski space.

Next, we take a position vector to be the prototype vector and use its transformation formula to define the general vector concept. Thus, if

$$\vec{R}_k = \begin{pmatrix} x_{1,k} \\ x_{2,k} \\ x_{3,k} \\ x_{4,k} \end{pmatrix}, \qquad \vec{R}_k' = \begin{pmatrix} x_{1,k}' \\ x_{2,k}' \\ x_{3,k}' \\ x_{4,k}' \end{pmatrix}, \qquad (8.15)$$

then under Lorentz transformation L, given by (6.19),

$$\vec{R}_k' = L\vec{R}_k \qquad (8.16)$$

Then, any quantity which has four components and is given in the lab frame by

$$\vec{W} = \begin{pmatrix} w_1 \\ w_2 \\ w_3 \\ w_4 \end{pmatrix}$$

and in the rocket frame by

$$\vec{W}' = \begin{pmatrix} w_1' \\ w_2' \\ w_3' \\ w_4' \end{pmatrix}$$

is called a 4-vector if and only if

$$\vec{W}' = L\vec{W}. \tag{8.17}$$

Finally, given two 4-vectors $\vec{W}^{(1)}$ and $\vec{W}^{(2)}$ in, say, the lab frame, we define the inner product $\vec{W}^{(1)} \cdot \vec{W}^{(2)}$ by

$$\vec{W}^{(1)} \cdot \vec{W}^{(2)} = w_1^{(1)} w_1^{(2)} + w_2^{(1)} w_2^{(2)} + w_3^{(1)} w_3^{(2)} + w_4^{(1)} w_4^{(2)}. \tag{8.18}$$

In developing special relativistic mechanics, we will proceed by using only Minkowski coordinates. No concepts in addition to those defined in this section will be required.

8.4 4-VELOCITY AND 4-ACCELERATION

Suppose now that particle P is in motion and in the lab frame is at (x_k, y_k, z_k) at time t_k while in the rocket frame it is at (x_k', y_k', z_k') at the corresponding time t_k'. At time t_k in the lab frame, we define P's Minkowski 4-velocity $\vec{V}_k$ and Minkowski 4-acceleration $\vec{A}_k$ by

$$\vec{V}_k = \begin{pmatrix} V_{1,k} \\ V_{2,k} \\ V_{3,k} \\ V_{4,k} \end{pmatrix} = \begin{pmatrix} \dfrac{\Delta x_{1,k}}{\delta \tau_k} \\ \dfrac{\Delta x_{2,k}}{\delta \tau_k} \\ \dfrac{\Delta x_{3,k}}{\delta \tau_k} \\ \dfrac{\Delta x_{4,k}}{\delta \tau_k} \end{pmatrix} ; \tag{8.19}$$

$$\vec{A}_k = \begin{pmatrix} A_{1,k} \\ \\ A_{2,k} \\ \\ A_{3,k} \\ \\ A_{4,k} \end{pmatrix} = \begin{pmatrix} \dfrac{\Delta V_{1,k}}{\delta \tau_k} \\ \\ \dfrac{\Delta V_{2,k}}{\delta \tau_k} \\ \\ \dfrac{\Delta V_{3,k}}{\delta \tau_k} \\ \\ \dfrac{\Delta V_{4,k}}{\delta \tau_k} \end{pmatrix} . \tag{8.20}$$

By the axiom of relativity, and recalling that $\delta \tau_k$ is invariant, $\vec{V}_k{}'$ and $\vec{A}_k{}'$ are defined in the rocket frame by

$$\vec{V}_k{}' = \begin{pmatrix} V_{1,k}{}' \\ \\ V_{2,k}{}' \\ \\ V_{3,k}{}' \\ \\ V_{4,k}{}' \end{pmatrix} = \begin{pmatrix} \dfrac{\Delta x_{1,k}{}'}{\delta \tau_k} \\ \\ \dfrac{\Delta x_{2,k}{}'}{\delta \tau_k} \\ \\ \dfrac{\Delta x_{3,k}{}'}{\delta \tau_k} \\ \\ \dfrac{\Delta x_{4,k}{}'}{\delta \tau_k} \end{pmatrix} ; \tag{8.19'}$$

$$\vec{A}_k{}' = \begin{pmatrix} A_{1,k}{}' \\ \\ A_{2,k}{}' \\ \\ A_{3,k}{}' \\ \\ A_{4,k}{}' \end{pmatrix} = \begin{pmatrix} \dfrac{\Delta V_{1,k}{}'}{\delta \tau_k} \\ \\ -\dfrac{\Delta V_{2,k}{}'}{\delta \tau_k} \\ \\ \dfrac{\Delta V_{3,k}{}'}{\delta \tau_k} \\ \\ \dfrac{\Delta V_{4,k}{}'}{\delta \tau_k} \end{pmatrix} . \tag{8.20'}$$

Direct computation reveals easily that both $\vec{V}_k$ and $\vec{A}_k$ are 4-vectors. The relationships between components of $\vec{v}_k$ in (8.1) and the first three components of $\vec{V}_k$ in (8.19) follow readily from (8.11) and (8.13). Similar connections can be established between $\vec{a}_k$ and the first three components of $\vec{A}_k$.

The magnitude V_k of 4-vector $\vec{V}_k$ is defined by

$$V_k = [-(\vec{V}_k \cdot \vec{V}_k)]^{\frac{1}{2}}.$$

An analogous definition holds for the magnitude V_k' of $\vec{V}_k'$. Then

$$V_k = \left[-\left(\frac{\Delta x_{1,k}^2 + \Delta x_{2,k}^2 + \Delta x_{3,k}^2 + \Delta x_{4,k}^2}{\delta \tau_k^2} \right) \right]^{\frac{1}{2}} \equiv 1. \tag{8.21}$$

In a completely similar fashion it follows that $V_k' \equiv 1$. Thus, the fact that all 4-velocities have magnitude unity indicates immediately that 4-velocity does not correspond to the usual concept of velocity.

Note finally that 4-vectors with respect to coordinates x_k, y_k, z_k and t_k can also be defined in a manner similar to that given above merely by replacing L with L^* in (8.17).

8.5 MOMENTUM AND ENERGY

We proceed now under the assumption that, without the presence of an external force, the interaction of two particles conserves linear momentum. To be precise, let particle P of mass $m(k)$ at time t_k be in motion in the lab frame. At time t_k, the linear momentum $\vec{p}_k$ of P is defined in the lab frame by

$$\vec{p}_k = m(k)\vec{v}_k \tag{8.22}$$

and in the rocket frame by

$$\vec{p}_k' = m'(k)\vec{v}_k', \tag{8.23}$$

where, as in Section 7.4,

$$m(k) = \frac{cm_0}{(c^2 - v_k^2)^{\frac{1}{2}}}, \tag{8.24}$$

$$m'(k) = \frac{cm_0}{(c^2 - v_k'^2)^{\frac{1}{2}}}, \tag{8.25}$$

and m_0 is called the rest mass of P.

The total energy E of particle P of mass $m(k)$ at time t_k is defined by

$$E = m(k)c^2. \tag{8.26}$$

Then, as in Section 7.6, the rest energy formula

$$E_0 = m_0 c^2 \tag{8.27}$$

follows readily, as do the formulas

$$E = m_0 c^3 \frac{\Delta t_k}{\delta \tau_k} \tag{8.28}$$

and

$$E^2 = p_k{}^2 c^2 + m_0{}^2 c^4, \tag{8.29}$$

where p_k is the magnitude of $\vec{p}_k$.

8.6   THE MOMENTUM-ENERGY 4-VECTOR

We proceed next to establish the momentum-energy results of Section 7.7.

Let

$$E^* = \frac{E}{c^2}. \tag{8.30}$$

We wish to show that

$$\begin{bmatrix} m(k)v_{1,k} \\ m(k)v_{2,k} \\ m(k)v_{3,k} \\ E^* \end{bmatrix}$$

is a 4-vector under L^*, called the momentum-energy vector. To do this, note that with the aid of (8.9), (8.24) and (8.25),

$$L^* \begin{bmatrix} m(k)v_{1,k} \\ m(k)v_{2,k} \\ m(k)v_{3,k} \\ E^* \end{bmatrix} = L^* \begin{bmatrix} m(k)v_{1,k} \\ m(k)v_{2,k} \\ m(k)v_{3,k} \\ m(k) \end{bmatrix} = L^* \begin{bmatrix} \frac{cm_0 \Delta x_k}{(c^2-v_k^2)^{1/2} \Delta t_k} \\ \frac{cm_0 \Delta y_k}{(c^2-v_k^2)^{1/2} \Delta t_k} \\ \frac{cm_0 \Delta z_k}{(c^2-v_k 2)^{1/2} \Delta t_k} \\ \frac{cm_0}{(c^2-v_k^2)^{1/2}} \end{bmatrix}$$

$$= L* \begin{bmatrix} \dfrac{cm_0}{\delta\tau_k} \Delta x_k \\[2ex] \dfrac{cm_0}{\delta\tau_k} \Delta y_k \\[2ex] \dfrac{cm_0}{\delta\tau_k} \Delta z_k \\[2ex] \dfrac{cm_0}{\delta\tau_k} \Delta t_k \end{bmatrix} = \begin{bmatrix} \dfrac{cm_0}{\delta\tau_k} \Delta x_k{'} \\[2ex] \dfrac{cm_0}{\delta\tau_k} \Delta y_k{'} \\[2ex] \dfrac{cm_0}{\delta\tau_k} \Delta z_k{'} \\[2ex] \dfrac{cm_0}{\delta\tau_k} \Delta t_k{'} \end{bmatrix} = \begin{bmatrix} m'(k)v_{1,k}{'} \\[2ex] m'(k)v_{2,k}{'} \\[2ex] m'(k)v_{3,k}{'} \\[2ex] m'(k) \end{bmatrix} ,$$

and the assertion is proved.

8.7 DYNAMICS

Finally, let us examine the development of a dynamical equation which is symmetric under the Lorentz transformation.

It is a most unfortunate mathematical and physical consequence of special relativity theory [7, p. 103] that the simple generalization

$$\vec{F} = \frac{d}{dt} (m\vec{v}) \tag{8.31}$$

of (7.24) does not, in general, transform under L* into

$$\vec{F}{'} = \frac{d}{dt{'}} (m'\vec{v}{'}) \tag{8.32}$$

even in the continuous case. To resolve this failure of the axiom of relativity with respect to (8.31), two approaches have been followed. First ([4, p. 268], [88, p. 63]), one might proceed under the approximating assumption that if a rocket frame were attached to P, so that it can have accelerated motion, and if at time t the velocity of P is $\vec{v}$, then one can treat the rocket frame at time t as being instantaneously in uniform relative motion with velocity $\vec{v}$ with respect to the lab frame. Indeed, such an assumption is tacitly made in the well known "clock paradox" [88, p. 63]. Second, ([7, p. 103], [92, p. 165]), one can formulate equations of motion directly in Minkowski space. Because the second approach has a firmer mathematical basis than does the first, we will explore it.

In Minkowski space, we assume the dynamical difference equation

$$\vec{F}_k = \alpha_k m(k)\vec{A}_k - \frac{\Delta(\alpha_k m(k))}{\delta\tau_k} \left(\frac{\vec{v}_{k+1} + \vec{v}_k}{2} \right) , \tag{8.33}$$

where

$$\alpha_k m(k) = m_0 . \tag{8.34}$$

We also define a three dimensional projection of (8.33) by

$$\vec{F}_k{}^P = c^2 [m(k)\vec{A}_k{}^P - \frac{\Delta m(k)}{\delta\tau_k} \vec{V}_k{}^P], \tag{8.35}$$

where c^2 and $\vec{V}_k{}^P$ have replaced α_k and $\frac{V_{k+1}+V_k}{2}$, respectively, in (8.33), and where the superscript p denotes the dropping of the fourth component of the given 4-vector.

Let us show first that (8.35) reduces to (7.23) when P and the rocket frame both have velocities in the X direction only. We must show then that

$$F_{1,k} = c^2 [m(k)A_{1,k} - \frac{\Delta m(k)}{\delta\tau_k} V_{1,k}] \tag{8.36}$$

is the same as (7.23). Now, from (8.19),

$$c^2 [m(k)A_{1,k} - \frac{\Delta m(k)}{\delta\tau_k} V_{1,k}] = \frac{c^2}{\delta\tau_k} [m(k)V_{1,k+1} - m(k+1)V_{1,k}]. \tag{8.37}$$

Note next that from (8.9) and (8.24)

$$m(k) = \frac{cm_0 \Delta t_k}{\delta\tau_k}, \tag{8.38}$$

so that (8.36)-(8.38) yield

$$F_{1,k} = \frac{cm(k)}{m_0 \Delta t_k} (m(k)V_{1,k+1} - m(k+1)V_{1,k})$$

$$= \frac{c^2 m(k)}{\frac{\delta\tau_k}{\Delta t_k} \cdot \frac{\delta\tau_{k+1}}{\Delta t_{k+1}}} \cdot \frac{\frac{\Delta x_{k+1}}{\delta\tau_{k+1}} \cdot \frac{\delta\tau_{k+1}}{\Delta t_{k+1}} - \frac{\Delta x_k}{\delta\tau_k} \cdot \frac{\delta\tau_k}{\Delta t_k}}{\Delta t_k},$$

which implies

$$F_{1,k} = \frac{c^2 m(k)}{[(c^2-v_k{}^2)(c^2-v_{k+1}{}^2)]^{1/2}} \cdot \frac{\Delta V_{1,k}}{\Delta t_k}. \tag{8.39}$$

But (8.39) is simply (7.23) in the new notation, and the result is proved.

In Minkowski space there is a problem in the study of (8.33), which will now be written as

$$\vec{F}_k = m_0 \vec{A}_k - \frac{\Delta m_0}{\delta\tau_k} \cdot \frac{\vec{V}_{k+1}+\vec{V}_k}{2}. \tag{8.40}$$

There is some question about whether or not rest mass should, in fact, depend on time through $\vec{F}_k$ ([4], [92]). In some sense this seems to undermine our physical intuition about what rest mass should be, but mathematically we could continue under this assumption as follows. If rest mass does depend on time through $\vec{F}_k$ then by taking inner products of both sides of (8.40) with $(V_{k+1}+V_k)/2$, and by using (8.19) and (8.21), one finds

$$\vec{F}_k \cdot \frac{\vec{v}_{k+1}+\vec{v}_k}{2} = -\frac{\Delta m_0}{\delta \tau_k}\left(\frac{\vec{v}_{k+1}+\vec{v}_k}{2}\right) \cdot \left(\frac{\vec{v}_{k+1}+\vec{v}_k}{2}\right) .$$

If then one restricts attention to forces $\vec{F}_k$ which satisfy

$$\vec{F}_k \cdot \frac{\vec{v}_{k+1}+\vec{v}_k}{2} = 0, \tag{8.41}$$

so that the force is normal to the average velocity, then one can always choose $\Delta m_0 = 0$, so that m_0 continues to be invariant. Restriction of attention to such forces also reduces (8.40) to

$$\vec{F}_k = m_0 \vec{A}_k , \tag{8.42}$$

which is invariant under the Lorentz transformation and is completely analogous in structure to Newton's equation of motion.

Note that, in the limit, (8.41) is the exact condition satisfied by the motion of a charged particle in an electromagnetic field.

Finally, with regard to (8.42), observe that for the special case $\vec{F}_k = \vec{0}$, one has

$$m_0 \frac{\vec{v}_{k+1}-\vec{v}_k}{\delta \tau_k} \equiv \vec{0},$$

so that

$$\vec{v}_k = \vec{v}_0, \quad k = 0,1,2,\dots . \tag{8.43}$$

But (8.43) implies

$$\frac{\vec{R}_{k+1}-\vec{R}_k}{\delta \tau_k} = \vec{v}_0, \quad k = 0,1,2,\dots ,$$

so that

$$\vec{R}_k = \vec{R}_0 + \vec{v}_0 \sum_{j=0}^{k-1} \delta \tau_j, \quad k = 1,2,3,\dots ,$$

which is a lineal trajectory in Minkowski space.

Chapter 9

Lorentz Invariant Computations

9.1 INTRODUCTION

When constructing a lab and a rocket frame, one is usually very careful to assume
identical clocks, identical units of length, and so forth, in each reference frame.
We assume, now, in addition, identical digital computers because we wish to analyze
the oscillatory motion of a particle P by means of such computers.

9.2 INVARIANT COMPUTATIONS

Our main emphasis in this chapter will be the arithmetic modeling and analysis of a
harmonic oscillator, typical of which is an electron in an atom [26]. Before we
consider such an oscillator, however, we will prove a fundamental result about the
related calculations, that is, that the computations in the lab and the rocket
frames will, in fact, be related by the Lorentz transformation.

As in Chapter 7, assume for the present that both the rocket frame and a given particle P are in motion in the X direction only. The motion of the particle in
the lab frame will be determined iteratively in the following way. Assuming that
x_k, t_k, v_k and F_k are given, then v_{k+1} is determined from (7.23), i.e., from

$$F_k = \frac{c^2 m(k)}{[(c^2-v_k^2)(c^2-v_{k+1}^2)]^{\frac{1}{2}}} \cdot \frac{\Delta v_k}{\Delta t_k} , \qquad (9.1)$$

x_{k+1} is determined from (7.11), that is, from

$$x_{k+1} = x_k + v_k \Delta t_k , \qquad (9.2)$$

and t_{k+1} is determined from

$$t_{k+1} = (k+1)\Delta t. \qquad (9.3)$$

In the rocket frame t_{k+1}' must correspond to t_{k+1} by the formula (7.1), i.e.,

136

$$t_{k+1}' = \frac{c^2 t_{k+1} - u x_{k+1}}{c(c^2-u^2)^{\frac{1}{2}}} \, , \tag{9.4}$$

while v_{k+1}' is determined from x_k', $\Delta t_k'$, v_k' and

$$F_k' = \frac{c^2 m'(k)}{[(c^2-v_k'^2)(c^2-v_{k+1}'^2)]^{\frac{1}{2}}} \cdot \frac{\Delta v_k'}{\Delta t_k'} \, . \tag{9.5}$$

Finally, x_{k+1}' is determined from (7.11'), i.e.,

$$x_{k+1}' = x_k' + v_k' \Delta t_k' \, . \tag{9.6}$$

When we say that our computations are Lorentz invariant, we mean that

$$x_{k+1}' = \frac{c(x_{k+1} - u t_{k+1})}{(c^2-u^2)^{\frac{1}{2}}} \tag{9.7}$$

and

$$v_{k+1}' = \frac{c^2(v_{k+1}-u)}{c^2-uv_{k+1}} \, . \tag{9.8}$$

Let us then first prove (9.7) and (9.8). For simplicity we consider only the case $k = 0$, since the general case follows in exactly the same way.

From (7.1), (7.16), (7.17), (9.2) and (9.6)

$$x_1' = x_0' + v_0' \Delta t_0'$$

$$= \frac{c(x_0 - u t_0)}{(c^2-u^2)^{\frac{1}{2}}} + \frac{c^2(v_0 - u)}{c^2-uv_0} \cdot \frac{(c^2 \Delta t_0 - u \Delta x_0)}{c(c^2-u^2)^{\frac{1}{2}}}$$

$$= \frac{c(c^2 x_1 - c^2 u t_1 + u^2 t_0 v_0 - u v_0 x_1 + u^2 x_1 - u^2 x_0)}{(c^2-u^2)^{\frac{1}{2}}(c^2-uv_0)}$$

$$= \frac{c(x_1 - u t_1)}{(c^2-u^2)^{\frac{1}{2}}} \, ,$$

which proves (9.7).

Next, by the invariance of (9.1) under the Lorentz transformation,

$$\frac{c^2 m(0)}{[(c^2-v_0^2)(c^2-v_1^2)]^{\frac{1}{2}}} \cdot \frac{\Delta v_0}{\Delta t_0} = \frac{c^2 m'(0)}{[(c^2-v_0'^2)(c^2-v_1'^2)]^{\frac{1}{2}}} \cdot \frac{\Delta v_0'}{\Delta t_0'} \, ,$$

which, by (7.16), (7.17), (7.21) and (7.21'), is equivalent to

$$\frac{v_1-v_0}{(c^2-v_0^2)(c^2-v_1^2)^{\frac{1}{2}}} = \frac{v_1'-\dfrac{c^2(v_0-u)}{c^2-uv_0}}{\left[c^2-\dfrac{c^4(v_0-u)^2}{(c^2-uv_0)^2}\right](c^2-v_1'^2)^{\frac{1}{2}}} \cdot \frac{c(c^2-u^2)^{\frac{1}{2}}}{c^2-uv_0},$$

or,

$$\frac{v_1-v_0}{(c^2-v_1^2)^{\frac{1}{2}}} = \frac{(v_1'c^2-v_1'uv_0-c^2v_0+c^2u)}{(c^2-u^2)^{\frac{1}{2}}(c^2-v_1'^2)^{\frac{1}{2}}} . \tag{9.9}$$

Now, instead of squaring (9.9), solving the resulting quadratic equation for v_1', and then eliminating the extraneous root, let us simply guess the solution to be

$$v_1' = \frac{c^2(v_1-u)}{c^2-uv_1} . \tag{9.10}$$

Direct substitution of (9.10) into (9.9) then verifies that (9.10) is the desired root, since

$$\frac{v_1'c^2-v_1'uv_0-c^2v_0+c^2u}{c(c^2-u^2)^{\frac{1}{2}}(c^2-v_1'^2)^{\frac{1}{2}}} = \frac{\dfrac{c^4(v_1-u)}{c^2-uv_1} - \dfrac{uv_0c^2(v_1-u)}{c^2-uv_1} - c^2v_0+c^2u}{c(c^2-u^2)^{\frac{1}{2}}\left[c^2-\dfrac{c^4(v_1-u)^2}{(c^2-uv_1)^2}\right]^{\frac{1}{2}}}$$

$$= \frac{v_1-v_0}{(c^2-v_1^2)^{\frac{1}{2}}},$$

which proves (9.8).

9.3 AN ARITHMETIC, NEWTONIAN HARMONIC OSCILLATOR

An oscillator is a particle whose motion is back and forth over all, or part, of a straight or curved path. The most well studied oscillator is the harmonic oscillator, whose motion is along a straight line, whose energy is conserved, and the force on which is directly proportional to its distance from fixed point, usually taken as the origin. In the spirit of Chapter 4, we will show first in this section how to formulate and analyze a Newtonian harmonic oscillator using only arithmetic. Then, we will formulate and analyze a relativistic harmonic oscillator and will also compare the two models.

An arithmetic, Newtonian harmonic oscillator is defined to be a particle whose motion is one dimensional, say along an X axis, and is determined by

$$\frac{v_{k+1}+v_k}{2} = \frac{x_{k+1}-x_k}{\Delta t} \tag{9.11}$$

$$a_k = \frac{v_{k+1}-v_k}{\Delta t} \tag{9.12}$$

$$F_k = ma_k , \tag{9.13}$$

where

$$F_k = -w\frac{x_{k+1}+x_k}{2} , \tag{9.14}$$

and where w is a positive constant.

Let us show first that the oscillator defined by (9.11)–(9.14) is conservative with respect to energy. To do this, let

$$W_n = \sum_{k=0}^{n-1} (x_{k+1}-x_k)F_k , \quad n = 1,2,3,\dots . \tag{9.15}$$

Then, (1.19) is valid since that result was independent of the force. Thus,

$$W_n = K_n - K_0 , \quad n = 1,2,3,\dots , \tag{9.16}$$

where K_n and K_0 are the kinetic energies at t_n and t_0, respectively.

Next, substitution of (9.14) into (9.16) yields

$$W_n = -\frac{1}{2}w \sum_{k=0}^{n-1} (x_{k+1}^2-x_k^2)$$

$$= -\frac{1}{2}w(x_n^2-x_0^2),$$

so that

$$W_n = -\frac{1}{2}wx_n^2 + \frac{1}{2}wx_0^2 .$$

Defining the potential energy of P at t_k by

$$V_k = \frac{1}{2}wx_k^2$$

yields

$$W_n = -V_n + V_0 , \quad n = 1,2,3,\dots . \tag{9.17}$$

Finally, elimination of W_n between (9.16) and (9.17) yields

$$K_n + V_n = K_0 + V_0 , \quad n = 1,2,3,\dots , \tag{9.18}$$

which is the law of conservation of energy.

Let us now turn to an analysis of harmonic motion, which is, of course, the motion of a harmonic oscillator. For simplicity, we will restrict attention to m = w = 1, so that (9.11)-(9.14) imply

$$\frac{v_{k+1}-v_k}{\Delta t} = - \frac{x_{k+1}+x_k}{2} \tag{9.19}$$

$$\frac{v_{k+1}+v_k}{2} = \frac{x_{k+1}-x_k}{\Delta t}, \tag{9.20}$$

or equivalently,

$$-v_{k+1} + v_k = \frac{\Delta t}{2}(x_{k+1}+x_k) \tag{9.21}$$

$$v_{k+1} + v_k = \frac{2}{\Delta t}(x_{k+1}-x_k). \tag{9.22}$$

Termwise addition of (9.21) and (9.22) yields

$$v_k = \frac{\Delta t^2+4}{4\Delta t}x_{k+1} + \frac{\Delta t^2-4}{4\Delta t}x_k. \tag{9.23}$$

But (9.23) is valid for all k, so that

$$v_{k+1} = \frac{\Delta t^2+4}{4\Delta t}x_{k+2} + \frac{\Delta t^2-4}{4\Delta t}x_{k+1}. \tag{9.24}$$

Termwise addition of (9.23) and (9.24) yields, together with (9.22),

$$(\Delta t^2+4)x_{k+2} + 2(\Delta t^2-4)x_{k+1} + (\Delta t^2+4)x_k = 0. \tag{9.25}$$

Setting

$$x_k = z^k, \quad z \neq 0 \tag{9.26}$$

implies from (9.25) that

$$(\Delta t^2+4)z^2 + 2(\Delta t^2-4)z + (\Delta t^2+4) = 0, \tag{9.27}$$

for which we find two complex conjugate roots z_1, z_2:

$$z_1 = \frac{4-\Delta t^2}{4+\Delta t^2} + i\frac{4\Delta t}{4+\Delta t^2}$$

$$z_2 = \frac{4-\Delta t^2}{4+\Delta t^2} - i\frac{4\Delta t}{4+\Delta t^2}.$$

Now, $|z_1| = |z_2| = 1$, so that in polar form

$$z_1 = \cos \Delta\theta + i \sin \Delta\theta,$$

$$z_2 = \cos \Delta\theta - i \sin \Delta\theta,$$

where

$$\Delta\theta = \cos^{-1} \frac{4-\Delta t^2}{4+\Delta t^2} . \tag{9.28}$$

Hence, from (9.26), there are two possible choices for x_k, i.e.,

$$x_{k,1} = \cos(k\Delta\theta) + i \sin(k\Delta\theta) \tag{9.29}$$

$$x_{k,2} = \cos(k\Delta\theta) - i \sin(k\Delta\theta).$$

If one calls the general solution

$$x_k = (a+bi)x_{k,1} + (a-bi)x_{k,2} , \tag{9.30}$$

where a and b are arbitrary real constants, then (9.29) and (9.30) yield

$$x_k = 2a \cos(k\Delta 0) - 2b \sin(k\Delta\theta), \quad k = 0,1,2,\ldots . \tag{9.31}$$

Thus, harmonic motion is described completely in $X - \theta$ space by (9.31), where we observe from (9.31) that it is periodic and that it is bounded, since a and b are determined completely by any given initial data.

9.4 AN ARITHMETIC, RELATIVISTIC HARMONIC OSCILLATOR

An arithmetic, relativistic harmonic oscillator is one which is defined in, say, the lab frame, by (9.1) with $F_k = -wx_k$. With this definition, let us adopt abso-lute units with $m_0 = c = 1$ and consider the particular oscillator for which $w = 1$. Then, from (9.1) and (9.2), the motion of the oscillator is given by:

$$x_{k+1} = x_k + (\Delta t)v_k \tag{9.32}$$

$$v_{k+1} = \frac{v_k-(\Delta t)x_k(1-v_k^2)^{3/2}[1+x_k^2\Delta t^2(1-v_k^2)]^{1/2}}{1+x_k^2\Delta t^2(1-v_k^2)^2} \tag{9.33}$$

The general solution of (9.32)-(9.33) cannot be constructed in the fashion des-cribed for (9.21)-(9.22). However, if one prescribes initial conditions, say,

$$x(0) = x_0 = 0, \quad v(0) = v_0 , \tag{9.34}$$

then (9.32)-(9.33) are simple iteration formulas from the computer point of view, and so the motion can be generated recursively from (9.32)-(9.34). In particular, this was done for 30000 time steps with $\Delta t = 0.0001$ for each of the cases $v_0 = 0.001, 0.01, 0.05, 0.1, 0.3, 0.5, 0.7, 0.9$. The total running time on the

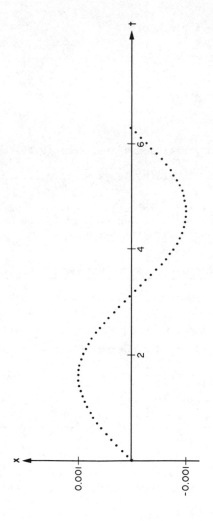

Fig. 9.1

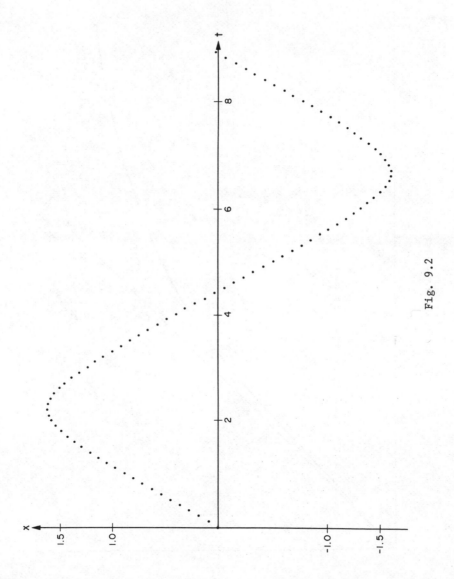

Fig. 9.2

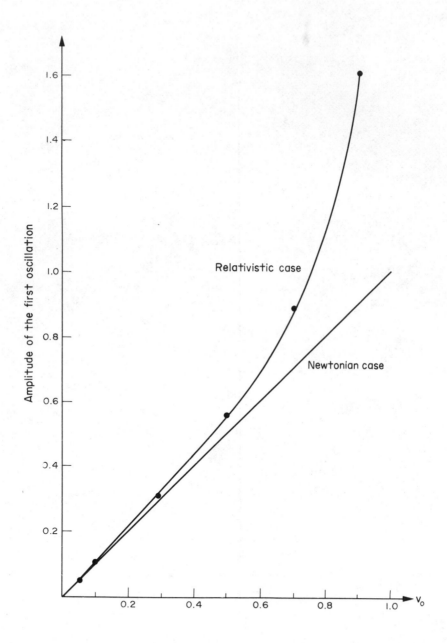

Fig. 9.3

UNIVAC 1110 was under two minutes. Figure 9.1 shows the amplitude and period of the first complete oscillation for the case $v_0 = 0.001$. For such a relatively low velocity, the oscillator should behave like a Newtonian oscillator and, indeed, this is the case, with the amplitude being 0.001 and, to two decimal places, the period being 6.28 ($\sim 2\pi$). Subsequent motion of this oscillator continues to show almost no change in amplitude or period. At the other extreme, Fig. 9.2 shows the motion for $v_0 = 0.9$, which is relatively close to the speed of light. To two decimal places, the amplitude of the first oscillation is 1.61 while the period is 8.88. These results are distinctly non-Newtonian, and, to thirty thousand time steps, these results remain constant to two decimal places but do show small increments in the third decimal place. Finally, in Fig. 9.3 is shown how the amplitude of a relativistic oscillator deviates from that of a Newtonian oscillator with increasing v_0.

9.5 MOTION OF AN ELECTRIC CHARGE IN A MAGNETIC FIELD

Finally, let us consider an example of motion in more than one dimension. Consider, in particular, the motion of an electric charge e, moving in the XY plane under the influence of a magnetic field which acts in the direction of the Z axis. Assume that in the XY plane the force on the charge is

$$\vec{F} = (eHv_y, -eHv_x), \tag{9.35}$$

where v is the speed of the charge and H is the intensity of the field. The relativistic differential equations of motion can be solved analytically to yield circular motion when H is uniform [92, p. 171]. When H is not uniform, then, in general, the equations of motion cannot be solved analytically.

In the spirit of our arithmetic formulations, let us introduce the absolute units $m_0 = c = e = 1$ so that

$$F_{k,x} = \frac{v_{k+1,x} - v_{k,x}}{\Delta t_k (1-v_k^2)(1-v_{k+1}^2)^{\frac{1}{2}}}, \quad F_{k,y} = \frac{v_{k+1,y} - v_{k,y}}{\Delta t_k (1-v_k^2)(1-v_{k+1}^2)^{\frac{1}{2}}}. \tag{9.36}$$

From (9.35) and (9.36), one has then

$$v_{k+1,x} - v_{k,x} - Hv_{k,y}(1-v_{k,x}^2-v_{k,y}^2)(1-v_{k+1,x}^2-v_{k+1,y}^2)^{\frac{1}{2}}\Delta t_k = 0, \tag{9.37}$$

$$v_{k+1,y} - v_{k,y} + Hv_{k,x}(1-v_{k,x}^2-v_{k,y}^2)(1-v_{k+1,x}^2-v_{k+1,y}^2)^{\frac{1}{2}}\Delta t_k = 0. \tag{9.38}$$

One can, of course, readily construct a related 4-force and a related set of Lorentz invariant dynamical equations, as shown in Chapter 8. However, the numerical computations are done most simply in cartesian space using (9.37), (9.38) and

$$x_{k+1} = x_k + v_{k,x}\Delta t_k$$
$$y_{k+1} = y_k + v_{k,y}\Delta t_k, \tag{9.39}$$

so, let us continue, in the present case, to concentrate on these. In particular, let us consider the initial conditions

$$x_0 = y_0 = v_{0,x} = 0, \quad v_{0,y} = 0.01. \tag{9.40}$$

For the parameter choices $\Delta t = 0.0001$ and $H = 100$, Fig. 9.4 shows the resulting circular trajectory T_1 with center at $(0.0001,0)$, radius $r = 0.0001$, and period $\tau = 0.063$, in complete agreement with the analytical solution [92, p. 171]. Equations (9.37) and (9.38) are solved at each time step by the generalized Newton's method with the velocity components at the previous time step being used to initiate the iteration.

Consider next the initial value problem defined by (9.36)-(9.40), but in a nonuniform magnetic field with a $1/r^2$ intensity given by

$$H = \frac{100}{1+\alpha(x^2+y^2)}, \quad \alpha \geq 0. \tag{9.41}$$

Of course, for $\alpha = 0$, (9.41) reduces to the above uniform case, where $H = 100$. For the parameter choices $\Delta t = 0.0001$ and $\alpha = 10^7$, the resulting particle trajectory T_2 is shown also in Fig. 9.4. The particle motion is initially similar to the circular motion of the first example, but as (x^2+y^2) increases and decreases, the varying effect of H results in the spiral type motion shown up to $t = 0.2$ in the figure.

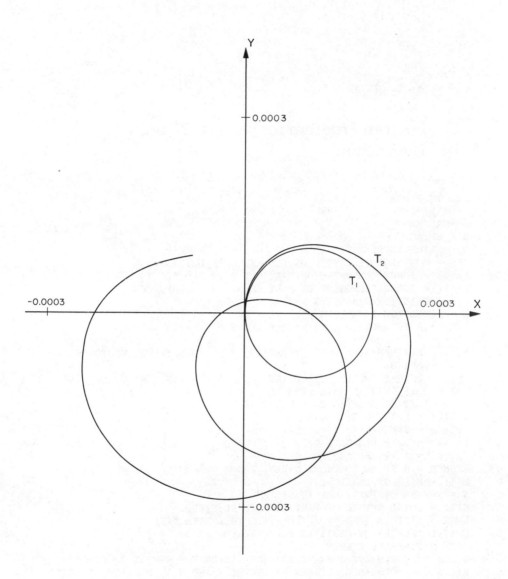

Fig. 9.4

Appendix 1

Fortran Program for General N-body Interaction

```
C  INDEX TO PROGRAM VARIABLES

C    A = ALPHA
C    B = BETA
C    DT = TIME INCREMENT
C    EPS = CONVERGENCE CRITERION FOR NEWTON'S METHOD
C    FX(I) = FORCE COMPONENT ON PARTICLE I IN X-DIRECTION
C    FY(I) = FORCE COMPONENT ON PARTICLE I IN Y-DIRECTION
C    G = ATTRACTION PARAMETER
C    H = REPULSION PARAMETER
C    IEND = 0 IF ANOTHER DATA CASE FOLLOWS
C         = 1 IF END OF RUN
C    IMAX = MAXIMUM NUMBER OF ITERATIONS PER TIMESTEP FOR NEWTON'S
C           METHOD
C    INIT = USER-SUPPLIED SUBROUTINE TO CALCULATE INITIAL CONDITIONS
C    INP = 0 IF INITIAL DATA READ IN
C        = 1 IF CALCULATED IN SUBROUTINE
C    IPRINT = PRINT-STEP INCREMENT
C    IPUNCH = PUNCH-STEP INCREMENT
C    ISTART = 0 IF NEW DATA CASE
C           = 1 IF RESTART
C    JPUNCH = 0 IF NO PUNCH REQUIRED, PUNCH OTHERWISE
C    M(I) = MASS OF PARTICLE I
C    N = NUMBER OF PARTICLES IN SYSTEM
C    NFIX = TOTAL NUMBER OF PARTICLES TO BE FIXED
C    NMAX = MAXIMUM NUMBER OF TIMESTEPS THIS DATA CASE
C    NP(I) = NUMBER OF PARTICLE TO BE FIXED
C    NSTP = TIMESTEP NUMBER
C    OMEGA = SUCCESSIVE OVER-RELAXATION FACTOR FOR NEWTON'S METHOD
C    R(I,J,1) = DISTANCE BETWEEN PARTICLES I AND J AT PREVIOUS TIMESTEP
C    R(I,J,2) = DISTANCE BETWEEN PARTICLES I AND J AT CURRENT TIMESTEP
C    VX(I,1) = X-COMPONENT OF VELOCITY OF PARTICLE I, PREVIOUS TIMESTEP
C    VX(I,2) = SAME AS ABOVE, CURRENT TIMESTEP, PREVIOUS ITERATION
C    VX(I,3) = SAME AS ABOVE, CURRENT TIMESTEP, CURRENT ITERATION
C    VX0(I) = X-COMPONENT OF INITIAL VELOCITY, PARTICLE I
C    VY(I,1) = Y-COMPONENT OF VELOCITY OF PARTICLE I,
C    VY(I,2) =      SAME DEFINITIONS
C    VY(I,3) =          AS VX(I,J), ABOVE
```

```
C   VY0(I) = Y-COMPONENT OF INITIAL VELOCITY, PARTICLE I
C   X(I,1) = X-COMPONENT OF POSITION OF PARTICLE I,
C   X(I,2) =        SAME DEFINITIONS
C   X(I,3) =          AS VX(I,J)
C   X0(I) = X-COMPONENT OF INITIAL POSITION, PARTICLE I
C   Y(I,1) = Y-COMPONENT OF POSITION OF PARTICLE I,
C   Y(I,2) =        SAME DEFINITIONS
C   Y(I,3) =          AS VX(I,J)
C   Y0(I) = Y-COMPONENT OF INITIAL POSITION, PARTICLE I
C
     IMPLICIT DOUBLE PRECISION(A-H,M,O-Z)
     DIMENSION X0(100),Y0(100),VX0(100),VY0(100),X(100,3),Y(100,3),
    1VX(100,3),VY(100,3),FX(100),FY(100),NP(100),R(45,45,2),M(100)

1001 FORMAT(8D10.0)
1002 FORMAT(18I5)
1003 FORMAT(4D10.0,I5)
1004 FORMAT(I5,2D10.0)
2000 FORMAT(1H1)
2001 FORMAT(5X,'N',5X,'OMEGA',5X,'EPS',5X,'IMAX',/,I7,F10.4,E8.1,I9)
2002 FORMAT(/6X,'A',8X,'B',8X,'G',8X,'H',7X,'DT'/4F9.3,E9.3)
2003 FORMAT(' TIMESTEP ',6X,'M',9X,'X',14X,'Y'/2X,I6)
2004 FORMAT(10X,F8.4,5F15.10)
2005 FORMAT('  NON-CONVERGENCE AFTER ',I3,' ITERATIONS FOR TIMESTEP= ',
    1I6,' DT= ',E8.2)
2501 FORMAT(2D25.18)

  10 PRINT 2000
     NSTP=0
     READ 1001,OMEGA,EPS,DT
     READ 1002,NMAX,IMAX,IPRINT,IPUNCH,JPUNCH,ISTART
     READ 1002,N,NFIX,INP
     READ 1003,A,B,G,H,IEND
     PRINT 2001,N,OMEGA,EPS,IMAX
     NM1=N-1
     IF(ISTART.EQ.0)GO TO 1
C  RESTART
     READ 1002,NSTP
     READ 2501,(X0(I),Y0(I),I=1,N)
     READ 2501,(VX0(I),VY0(I),I=1,N)
     GO TO 3
C  NEW CASE
   1 IF(INP.NE.0)GO TO 2
     READ 1001,(X0(I),Y0(I),I=1,N),(VX0(I),VY0(I),I=1,N)
     GO TO 3
   2 CALL INIT
   3 IF(NFIX.NE.0)READ 1002,(NP(I),I=1,NFIX)
     READ 1001,(M(I),I=1,N)
     PRINT 2002,A,B,G,H,DT
     OMW=1.0-OMEGA
     T=0.0
     DT2=DT/2.0
     PRINT 2003,NSTP
     DO 30 I=1,N
     PRINT 2004,M(I),X0(I),Y0(I)
  30 CONTINUE
```

```
C   SPECIFY INITIAL GUESS FOR NEWTON'S ITERATION AT FIRST TIMESTEP
        DO 40 I=1,N
        X(I,3)=X0(I)
        VX(I,3)=VX0(I)
        Y(I,3)=Y0(I)
    40 VY(I,3)=VY0(I)
        CALL RCALC
C   UPDATE POSITIONS,VELOCITIES,DISTANCES FOR ALL TIMESTEPS
    45 NSTP=NSTP+1
        T=T+DT
        DO 60 I=1,N
        X(I,1)=X(I,3)
        VX(I,1)=VX(I,3)
        Y(I,1)=Y(I,3)
        VY(I,1)=VY(I,3)
        DO 50 J=1,N
    50 R(I,J,1)=R(I,J,2)
    60 CONTINUE

C   BEGIN ITERATION LOOP
        DO 90 K=1,IMAX
C   UPDATE ALL VARIABLES, CURRENT TIMESTEP, PREVIOUS ITERATION
        DO 70 I=1,N
        X(I,2)=X(I,3)
        VX(I,2)=VX(I,3)
        Y(I,2)=Y(I,3)
        VY(I,2)=VY(I,3)
    70 CONTINUE
C   UPDATE POSITIONS, CURRENT TIMESTEP, CURRENT ITERATION
        DO 73 I=1,N
        IF(NFIX.EQ.0)GO TO 72
        DO 71 J=1,NFIX
        IF(I.EQ.NP(J))GO TO 73
    71 CONTINUE
    72 X(I,3)=OMW*X(I,2)+OMEGA*(DT2*(VX(I,2)+VX(I,1))+X(I,1))
        Y(I,3)=OMW*Y(I,2)+OMEGA*(DT2*(VY(I,2)+VY(I,1))+Y(I,1))
    73 CONTINUE
        CALL RCALC
        CALL FCALC
C   UPDATE VELOCITIES, CURRENT TIMESTEP, CURRENT ITERATION
        DO 80 I=1,N
        IF(NFIX.EQ.0)GO TO 75
        DO 74 J=1,NFIX
        IF(I.EQ.NP(J))GO TO 80
    74 CONTINUE
    75 VX(I,3)=OMW*VX(I,2)+OMEGA*(DT*FX(I)+VX(I,1))
        VY(I,3)=OMW*VY(I,2)+OMEGA*(DT*FY(I)+VY(I,1))
    80 CONTINUE
C   TEST FOR CONVERGENCE
        DO 85 I=1,N
        IF(ABS(X(I,3)-X(I,2)).GT.EPS)GO TO 90
        IF(ABS(Y(I,3)-Y(I,2)).GT.EPS)GO TO 90
        IF(ABS(VX(I,3)-VX(I,2)).GT.EPS)GO TO 90
        IF(ABS(VY(I,3)-VY(I,2)).GT.EPS)GO TO 90
    85 CONTINUE
        GO TO 95
    90 CONTINUE
```

```
       PRINT 2005,K,NSTP
       GO TO 110
   95 IF(MOD(NSTP,IPRINT).NE.0)GO TO 103
       IF(JPUNCH.EQ.0)GO TO 102
       IF(MOD(NSTP,IPUNCH).NE.0)GO TO 102
       WRITE(1,1002)NSTP
       DO 151 I=1,N
       WRITE(1,2501)X(I,3),Y(I,3)
  151 CONTINUE
       DO 152 I=1,N
       WRITE(1,2501)VX(I,3),VY(I,3)
  152 CONTINUE
       IF(NFIX.NE.0)WRITE(1,1002)(NP(II),II=1,NFIX)
  102 CALL OUTP
  103 IF(NSTP.EQ.NMAX)GO TO 110
       GO TO 45
  110 IF(IEND.EQ.0)GO TO 10
       STOP

C  INTERNAL SUBROUTINE TO COMPUTE DISTANCES BETWEEN PARTICLES
       SUBROUTINE RCALC
       DO 210 I=1,NM1
       IP1=I+1
       DO 200 J=IP1,N
       R(I,J,2)=DSQRT((X(I,3)-X(J,3))**2+(Y(I,3)-Y(J,3))**2)
       R(J,I,2)=R(I,J,2)
  200 CONTINUE
  210 CONTINUE
       RETURN

C  INTERNAL SUBROUTINE TO COMPUTE FORCES (ACCELERATIONS)
       SUBROUTINE FCALC
       IA=A-1
       IB=B-1
       DO 600 I=1,N
       IF(NFIX.EQ.0)GO TO 450
       DO 400 K=1,NFIX
       IF(I.EQ.NP(K))GO TO 600
  400 CONTINUE
  450 SUMX=0.
       SUMY=0.
       DO 550 J=1,N
       IF(I.EQ.J)GO TO 550
       SUMG=0.
       SUMH=0.
       RIJ=R(I,J,1)+R(I,J,2)
       DO 500 I2=1,IA
       SUMG=SUMG+(R(I,J,1)**(I2-1))*(R(I,J,2)**(IA-I2))
  500 CONTINUE
       DO 501 I2=1,IB
       SUMH=SUMH+(R(I,J,1)**(I2-1))*(R(I,J,2)**(IB-I2))
  501 CONTINUE
       GD=(R(I,J,1)*R(I,J,2))**IA*RIJ
       SUMG=G*SUMG/GD
       HD=(R(I,J,1)*R(I,J,2))**IB*RIJ
       SUMH=H*SUMH/HD
       SUMX=SUMX+(SUMH-SUMG)*M(J)*(X(I,3)+X(I,1)-X(J,3)-X(J,1))
       SUMY=SUMY+(SUMH-SUMG)*M(J)*(Y(I,3)+Y(I,1)-Y(J,3)-Y(J,1))
```

```
  550 CONTINUE
      FX(I)=SUMX
      FY(I)=SUMY
  600 CONTINUE
      RETURN

C  INTERNAL PRINT SUBROUTINE
      SUBROUTINE OUTP
 3001 FORMAT(2X,IS)
 3002 FORMAT(15X,5F15.10)
      PRINT 3001,NSTP
      DO 800 II=1,N
      PRINT 3002,X(II,3),Y(II,3)
  800 CONTINUE
      RETURN


C  INTERNAL SUBROUTINE TO CALCULATE INITIAL CONDITIONS
C  FOR DISCRETE CONDUCTIVE HEAT TRANSFER
C     ANGLE(I) = ANGLE (IN DEGREES) OF INITIAL VELOCITY VECTOR WITH
C                RESPECT TO POSITIVE X-AXIS
C     IAXIS(I) = 0 IF LEFTMOST PARTICLE IN ROW I IS ON Y-AXIS
C              = +1,-1 IF LEFTMOST PARTICLE TO BE SHIFTED RIGHT OR LEFT,
C                RESPECTIVELY
C     IROW(I) = NUMBER OF PARTICLES IN ROW I
C     IVEL(I) = NUMBER OF PARTICLE TO BE GIVEN AN INITIAL VELOCITY
C     NAXIS = NUMBER OF ROW ON X-AXIS
C     NROW = NUMBER OF ROWS IN SYSTEM
C     NVEL = TOTAL NUMBER OF PARTICLES TO BE GIVEN AN INITIAL VELOCITY
C     VEL(I) = MAGNITUDE OF INITIAL VELOCITY VECTOR

      DIMENSION IROW(5),IAXIS(5),IVEL(5),VEL(5),ANGLE(5)
 4001 FORMAT(18I5)
 4002 FORMAT(I5,2D10.0)

C  CALCULATE POSITIONS
      BASEX=(H*(B-1)/(G*(A-1)))**(1.0/(B-A))
      BASEY=SQRT(BASEX**2-(0.5*BASEX)**2)
      READ 4001,(IROW(I),IAXIS(I),I=1,NROW)
      IL=1
      IU=0
      DO 3 I=1,MROW
      IU=IU+IROW(I)
      XSHIFT=0.5*IAXIS(I)*BASEX
      JSHIFT=NAXIS-I
      DO 2 J=IL,IU
      X0(J)=(J-IL)*BASEX+XSHIFT
      Y0(J)=JSHIFT*BASEY
    2 CONTINUE
      IL=IU+1
    3 CONTINUE
C  CALCULATE VELOCITIES
   10 DO 4 I=1,N
      VX0(I)=0.0
      VY0(I)=0.0
    4 CONTINUE
      IF(NVEL.EQ.0)RETURN
      READ 4002,(IVEL(I),VEL(I),ANGLE(I),I=1,NVEL)
```

```
      PI=3.14159265358979324D+00
      RAD=PI/180.0
      DO 5 I=1,NVEL
      J=IVEL(I)
      THETA=ANGLE(I)*RAD
      VX0(J)=VEL(I)*COS(THETA)
      VY0(J)=VEL(I)*SIN(THETA)
    5 CONTINUE
      RETURN
      END
```

Appendix 2

Fortran Program for Planetary-type Evolution

```
C FORTRAN PROGRAM FOR THE CREATION OF THE EARTH.
C RESULTS TO BE PUBLISHED IN OSSERVATORE ROMANO, SUNDAY EDITION.
      PARAMETER N=239,N2=N*2
C DIMENSION STATEMENT
      DIMENSION PMASS(N),XO(N),YO(N),VXO(N),VYO(N),X(N,2),Y(N,2),
     1VX(N,2),VY(N,2),ACX(N),ACY(N),TEMP(N),XREL(N),YREL(N),DIST(N),
     1VTREL(N),VNREL(N),VXREL(N),VYREL(N)
C  VARIABLES XO, YO,ETC., HAVE BEEN PUNCHED OH, NOT ZERO
C INITIALIZATION SET TO ZERO AUTOMATICALLY ON 1110
C WE NEXT ASSURE THAT X, Y, VX, VY, ACX, ACY ARE JUXTAPOSED WITHIN CORE.
      COMMON/DG/X,Y,VX,VY,ACX,ACY
      NM1=N-1
      K=1
      KDAMP=50
      KPRINT=100
C INITIAL DATA INPUT
      READ 10, (PMASS(I),XO(I),YO(I),VXO(I),VYO(I),TEMP(I),I=1,N)
10      FORMAT (5F10.3,F10.1)
C PRINT INITIAL DATA
      DO 20 I=1,N
      PRINT 15,I,PMASS(I),XO(I),YO(I),VXO(I),VYO(I),TEMP(I)
15      FORMAT (5X,I7,5F12.5,F12.1)
20      CONTINUE
C UPDATE. X(I,1) IS X COORDINATE AT PREVIOUS TIME STEP.  X(I,2) IS X
C  COORDINATE AT PRESENT TIME STEP.  SIMILARLY FOR OTHER VARIABLES.
C INITIALLY WE WOULD LIKE TO DO THE FOLLOWING FOR I=1,N
C     X(I,1)=XO(I)
C     Y(I,1)=YO(I)
C     VX(I,1)=VXO(I)
C     VY(I,1)=VYO(I)
C INSTEAD WE DO IT BY THE SUBROUTINE AS FOLLOWS.
      CALL COPVAR (XO(1),1,X(1,1),1,N)
      CALL COPVAR (YO(1),1,Y(1,1),1,N)
      CALL COPVAR (VXO(1),1,VX(1,1),1,N)
      CALL COPVAR (VYO(1),1,VY(1,1),1,N)
      GO TO 71
C WE WILL WANT TO UPDATE
C     X(I,1)=X(I,2)
```

```
C      Y(I,1)=Y(I,2)
C      VX(I,1)=VX(I,2)
C      VY(I,1)=VY(I,2)
C WHICH IS ACCOMPLISHED BY
65        CALL COPVAR (X(1,2),1,X(1,1),1,N)
          CALL COPVAR (Y(1,2),1,Y(1,1),1,N)
          CALL COPVAR (VX(1,2),1,VX(1,1),1,N)
          CALL COPVAR (VY(1,2),1,VY(1,1),1,N)
C CALCULATION OF ACCELERATIONS IS DONE THROUGH STEP 770
C      DO 701 I=1,N
C      ACX(I)=0.
C      ACY(I)=0.
C 701     CONTINUE
C ARE ACCOMPLISHED BY
          CALL COPCON (0.,0,ACX(1),1,N)
          CALL COPCON (0.,0,ACY(1),1,N)
71        DO 77 I=1,NM1
          IP1=I+1
          DO 76 J=IP1,N
          R2=(X(I,1)-X(J,1))**2+(Y(I,1)-Y(J,1))**2
          R=SQRT(R2)
          IF (R2.GT.5.29) GO TO 72
          F=5.*PMASS(I)*PMASS(J)/(R**6)-5.*PMASS(I)*PMASS(J)/(R**4)
          FX=F*(X(I,1)-X(J,1))/R
          FY=F*(Y(I,1)-Y(J,1))/R
          GO TO 75
72        F=-0.001*PMASS(I)*PMASS(J)/(R2)
          FX=F*(X(I,1)-X(J,1))/R
          FY=F*(Y(I,1)-Y(J,1))/R
C ACCUMULATION OF FORCES ON PARTICLE I DUE TO ALL OTHER PARTICLES IS DONE
C IN NEXT FOUR FORMULAS
75        ACX(I)=ACX(I)+FX
          ACX(J)=ACX(J)-FX
          ACY(I)=ACY(I)+FY
          ACY(J)=ACY(J)-FY
76        CONTINUE
77        CONTINUE
C NOTE THAT WE HAVE JUST ACCUMULATED FORCES, NOT ACCELERATIONS - THE ABOVE
C NOTATION, THOUGH MISLEADING, ENABLES US TO SAVE MEMORY LOCATIONS.  WE
C NEXT CALCULATE THE VELOCITIES AND POSITIONS AS FOLLOWS.
C ACTUAL ACCELERATIONS ARE INCLUDED BY DIVISION BY MASS OF 75 DIRECTLY.
C LEAPFROG FORMULAS ARE USED BUT SPECIAL STARTERS ARE NOT SHOWN BECAUSE
C PROGRAM HERE CONTINUES CALCULATIONS FROM PUNCHED OUTPUT.
79        DO 799 I=1,N
          VX(I,2)=VX(I,1)+.0001*ACX(I)/PMASS(I)
          VY(I,2)=VY(I,1)+.0001*ACY(I)/PMASS(I)
          X(I,2)=X(I,1)+.0001*VX(I,2)
          Y(I,2)=Y(I,1)+.0001*VY(I,2)
799       CONTINUE
800       K=K+1
C CALCULATE MASS CENTER AND AVERAGE VELOCITY
          XBAR=0.
          YBAR=0.
          VXBAR=0.
          VYBAR=0.
          DO 8003 I=1,N
          XBAR=XBAR+PMASS(I)*X(I,2)
          YBAR=YBAR+PMASS(I)*Y(I,2)
```

```
      VXBAR=VXBAR+PMASS(I)*VX(I,2)
      VYBAR=VYBAR+PMASS(I)*VY(I,2)
8003  CONTINUE
      XBAR=XBAR/618000.
      YBAR=YBAR/618000.
      VXBAR=VXBAR/618000.
      VYBAR=VYBAR/618000.
      PRINT 8006,XBAR,YBAR,VXBAR,VYBAR
8006     FORMAT (5X,4F10.2)
C FINALLY WE CALCULATE TEMPERATURE. THIS IS DONE RELATIVE TO (XBAR,YBAR).
C WE TAKE OUT ALL LINEAR AND ROTATIONAL SYSTEM VELOCITIES FROM EACH PARTICLE.
C WHAT REMAIN ARE THE INTERPARTICLE MOTIONS, WHICH DEFINE PARTICLE HEAT.
C WE FIRST SHIFT TO THE CENTER OF MASS.
      DO 8020 I=1,N
      XREL(I)=X(I,2)-XBAR
      YREL(I)=Y(I,2)-YBAR
      VXREL(I)=VX(I,2)-VXBAR
      VYREL(I)=VY(I,2)-VYBAR
8020  CONTINUE
C WE NEXT INTRODUCE TANG AND NORM VELOCITY COMPONENTS.
      DO 8022 I=1,N
      DIST(I)=SQRT(XREL(I)**2+YREL(I)**2)
      VTREL(I)=(-VXREL(I)*YREL(I)+VYREL(I)*XREL(I))/DIST(I)
      VNREL(I)=(VYREL(I)*YREL(I)+VXREL(I)*XREL(I))/DIST(I)
8022  CONTINUE
C OUT OF VTREL(I) TAKE THE AVERAGE ANGULAR VELOCITY OF THE SYSTEM.
C THIS AVERAGE ANGULAR VELOCITY IS THDOTB.
      THDOTB=0.
      DO 8024 I=1,N
      THDOTB=THDOTB+PMASS(I)*VTREL(I)/DIST(I)
8024  CONTINUE
      THDOTB=THDOTB/548000.
      PRINT 8026,THDOTB
8026  FORMAT (5X,F10.5)
C WE NOW SUBTRACT OFF THE SYSTEM ROTATIONAL VELOCITY.
      DO 8028 I=1,N
      VTREL(I)=VTREL(I)-DIST(I)*THDOTB
8028     CONTINUE
C FINALLY WE CALCULATE TEMP.  IT IS NORMALIZED BY DIVISION BY 1000 AND
C IS TAKEN AS THE AVERAGE OVER 500 TIME STEPS TO INDICATE THAT IT IS AN
C OBSERVED QUANTITY.  IT IS ESSENTIALLY KE.
      DO 8029 I=1,N
      TEM=VTREL(I)**2+VNREL(I)**2
      TEM=.0005*PMASS(I)*TEM
      TEMP(I)=(499.*TEMP(I)+TEM)/500.
8029     CONTINUE
C RADIATE , IN AND OUT, EVERY KDAMP STEPS.
      IF (MOD(K,KDAMP).GT.0) GO TO 8030
C WE TAKE ACCOUNT OF RADIATION FOR ALL OUTER PARTICLES, WHICH MEANS
C ALL PARTICLES A DISTANCE GREATER THAN 7 FROM THE CENTROID, BY ADDING
C OR SUBTRACTING FROM THEIR SPEEDS.  WE WORK WITH SQUARES OF SPEEDS IN
C ORDER TO ECONOMIZE ON THE SQUARE ROOT PROCEDURE.
C ASSUME THE SUN IS ABOVE THE CENTROID AND DARKNESS BELOW.
      DO 8018 I=1,N
      DISTM=(X(I,2)-XBAR)**2+(Y(I,2)-YBAR)**2
      IF (DISTM,LT,49.) GO TO 8018
      IF (Y(I,2).GT.YBAR) GO TO 8012
      VX(I,2)=0.9955*VX(I,2)
```

```
       VY(I,2)=0.9955*VY(I,2)
       GO TO 8018
8012      VX(I,2)=1.001*VX(I,2)
       VY(I,2)=1.001*VX(I,2)
8018      CONTINUE
8030      IF (MOD(K,KPRINT).GT.0) GO TO 82
       DO 810 I=1,N
        PRINT 81,K,I,X(I,2),Y(1,2),VX(I,2),VY(I,2),TEMP(I)
81        FORMAT (5X,2I7,5F12.4)
810       CONTINUE
C TERMINATION AFTER A FIXED NUMBER OF STEPS
82       IF (K.LT.500) GO TO 65
C PUNCH OUTPUT FOR RESTART
       PUNCH 10, (PMASS(I),X(I,2),Y(I,2),VX(I,2),VY(I,2),TEMP(I),I=1,N)
       STOP
       END
```

References and Sources for Further Reading

1. Adams, W. M. (Editor), (1970). <u>Tsunamis in the Pacific Ocean</u>. East-West Center Press, Honolulu.
2. Aguire-Puente, J., and M. Fremond (1976). Frost propagation in wet porous media. <u>Proceedings of Joint IUTAM/IMU Symposium on Application of Methods of Functional Analysis to Problems of Mechanics</u>, Marseille, 1975. Springer-Verlag, Berlin.
3. Alder, B. J. (1964). Studies in molecular dynamics. III: A mixture of hard spheres. <u>J. Chem. Phys.</u>, <u>40</u>, 2724-2730.
4. Arzelies, H. (1966). <u>Relativistic Kinematics</u>. Pergamon Press, New York.
5. Barker, J. A., and D. Hendersen (1976). What is 'liquid'? Understanding the states of matter. <u>Rev. Mod. Phys.</u>, <u>48</u>, 587-671.
6. Barto, A. G. (1975). <u>Cellular Automata as Models of Natural Systems</u>. Ph.D. Thesis, Univ. Michigan, Ann Arbor.
7. Bergmann, P. (1942). <u>Introduction to the Theory of Relativity</u>. Prentice-Hall, Englewood Cliffs, New Jersey.
8. Bilaniuk, O. M. P., V. K. Deshpande, and E. C. G. Sudarshan. (1962). 'Meta' Relativity. <u>Amer. J. Phys.</u>, <u>30</u>, 718-723.
9. Birkhoff, G., and R. E. Lynch (1961). Lagrangian hydrodynamic computations and molecular models of matter. <u>LA-2618</u>. Los Alamos Sci. Lab., New Mexico.
10. Bonnerot, R., and P. Jamet (1974). A second order finite element method for the one-dimensional Stefan problem. <u>Int. J. Num. Meth. Eng.</u>, <u>8</u>, 811-820.
11. Burggraf, O. R. (1966). Analytical and numerical studies of the structure of separated flows. <u>J. Fluid Mech.</u>, <u>24</u>, 113-151.
12. Cadzow, J. A. (1970). Discrete calculus of variations. <u>Int. J. Control</u>, <u>11</u>, 393-407.
13. Chorin, A. J. (1973). Numerical study of slightly viscous flow. <u>J. Fluid Mech.</u>, <u>57</u>, 785-796.
14. Ciavaldini, J. F. (1975). Analyse numerique d'un probleme de Stefan. <u>SIAM J. Num. Anal.</u>, <u>12</u>, 464-487.
15. Costabel, P. (1973). <u>Leibniz and Dynamics</u>. Cornell Univ. Press, Ithaca, New York.
16. Courant, R., and K. O. Friedrichs (1963). <u>Supersonic Flow and Shock Waves</u>. Interscience, New York.
17. Courant, R., and D. Hilbert (1953). <u>Methods of Mathematical Physics</u>, Vol. I. Interscience, New York.
18. Crank, J. (1957). Two methods for the numerical solution of moving-boundary problems in diffusion and heat flow. <u>QJMAM</u>, <u>10</u>, 220-231.

19. Cryer, C. W. (1969). Stability Analysis in Discrete Mechanics. TR #67.
     Computer Sciences Department, University of Wisconsin, Madison, Wisconsin.
20. Deeter, C. R. (1972). Discrete generalized functions. J. Math. Anal. Appl.,
     39, 375-396.
21. Douglas, J., and T. M. Gallie (1955). On the numerical integration of a para-
     bolic differential equation subject to a moving boundary condition. Duke
     Math. J., 122, 557-571.
22. Ehlers, J., K. Hepp, and H. A. Weidinmuller (Editors), (1971). Proceedings
     of the Second International Conference on Numerical Methods in Fluid
     Dynamics. Springer-Verlag, New York.
23. Ehrlich, L. W. (1958). A numerical method of solving a heat flow problem with
     moving boundary. J. ACM, 5, 161-176.
24. Einstein, A. (1905). Zur Elektrodynamik bewegter Korper. Annalen der Physik,
     4, 891-921.
25. Fermi, E., J. Pasta, and S. Ulam (1955). Studies of Nonlinear Problems I.
     LA-1940, Los Alamos Sci. Labs.
26. Feynman, R. P., R. B. Leighton, and M. Sands (1963). The Feynman Lectures on
     Physics. Addison-Wesley, Reading, Mass.
27. Forsythe, A. I., T. A. Keenan, E. I. Organick, and W. Stenberg (1969). Com-
     puter Science: A First Course. Wiley, New York.
28. Friedman, A. (1964). Partial Differential Equations of Parabolic Type.
     Prentice-Hall, Englewood Cliffs, New Jersey.
29. Greenspan, D. (1972). Discrete Newtonian gravitation and the N-body problem.
     Utilitas Math., 2, 105-126.
30. Greenspan, D. (1972). An energy conserving, stable discretization of the har-
     monic oscillator. Bull. Poly. Inst. Iasi, XVII (XXII), Section 1, 205-
     209.
31. Greenspan, D. (1973). Discrete Models. Addison-Wesley, Reading, Mass.
32. Greenspan, D. (1973). Symmetry in discrete mechanics. Foundations of Physics,
     3, 247-253.
33. Greenspan, D. (1974). Discrete Newtonian gravitation and the three-body prob-
     lem. Foundations of Physics, 4, 299-310.
34. Greenspan, D. (1974). Discrete bars, conductive heat transfer, and elasticity,
     Computers and Structures, 4, 243-251.
35. Greenspan, D. (1974). A physically consistent, discrete N-body model. Bull.
     Amer. Math. Soc., 80, 553-555.
36. Greenspan, D. (1974). An arithmetic, particle theory of fluid dynamics. Comp.
     Meth. Appl. Mech. Eng., 3, 293-303.
37. Greenspan, D. (1974). Discrete Numerical Methods in Physics and Engineering.
     Academic Press, New York.
38. Greenspan, D. (1978). Computer studies of a von Neuman type fluid. Appl.
     Math. and Comp., 4, 15-25.
39. Greenspan, D. (1975). Computer Newtonian and special relativistic mechanics.
     Proceedings of the Second USA-Japan Computer Conference, Amer. Fed. Inf.
     Proc. Soc., 88-91. Montvale, New Jersey.
40. Greenspan, D. (1976). The arithmetic basis of special relativity. Int. J.
     Theor. Phys., 15, 557-574.
41. Greenspan, D. (1976). Cavity flow of a particle fluid. Tech. Rpt. 265.
     Dept. Comp. Sci., Univ. Wisconsin, Madison, Wisconsin. Appears, in part,
     in Proceedings of Algorithms '79, Bratislava.
42. Greenspan, D. (1978). New mathematical models of porous flow. Tech. Rpt. 95,
     Dept. Math., Univ. Texas at Arlington.
43. Greenspan, D. (1978). N-body modeling of nonlinear, free surface liquid flow.
     Tech. Rpt. 97, Dept. Math., Univ. Texas at Arlington.
44. Greenspan, D., and J. Collier (1978). Computer studies of swirling particle
     fluids and the evolution of planetary-type bodies. JIMA, 22, 235-253.
45. Greenspan, D., M. Cranmer, and J. Collier (1976). A particle model of ocean
     waves generated by earthquakes. Tech. Rpt. 277, Dept. Comp. Sci., Univ.
     Wisconsin, Madison, Wisconsin.

46. Greenspan, D., and M. Rosati (1978). Computer generation of particle solids. Computers and Structures, 8, 107–112.

47. Greenspan, H. P. (1968). The Theory of Rotating Fluids. Cambridge University Press, Cambridge.

48. Harlow, F. H., and J. E. Welch (1965). Numerical calculation of time-dependent viscous, incompressible flow. Phys. Fluids, 8, 2182–2189.

49. Hart, P. J. (Editor), (1969). The Earth's Crust and Upper Mantle. Amer. Geo. Union, Washington, D.C.

50. Hirschfelder, J. O., C. F. Curtiss, and R. B. Bird (1954). Molecular Theory of Gases and Liquids. Wiley, New York.

51. Jamet, P., and R. Bonnerot (1973). Numerical computation of the free boundary for the two-dimensional Stefan problem by finite elements. Centre d'Etudes de Limeil, Service M.A., B.P. 27, 94190 Villeneuve-St-Georges, France. (To appear in Int. J. Num. Meth. Eng.)

52. Jamet, P., P. Lascaux, and P.-A. Raviart (1970). Une methode de resolution numerique des equations de Navier-Stokes. Numer. Math., 16, 93–114.

53. Kardestuncer, H. (1975). Discrete Mechanics — A Unified Approach. Springer-Verlag, New York.

54. Kawaguti, M. (1961). Numerical solution of the Navier-Stokes equations for the flow in a two-dimensional cavity. J. Phys. Soc. Japan, 16, 2307–2318.

55. Kneser, A. (1911). Die Integralgleichungen und ihre Anwendung in der Mathematischen Physik. Viewig & Sohn, Braunschweig, Germany.

56. LaBudde, R. A., and D. Greenspan (1974). Discrete mechanics — a general treatment. J. Comp. Physics, 15, 134–167.

57. LaBudde, R. A., and D. Greenspan (1976). Energy and momentum conserving methods of arbitrary order for the numerical integration of equations of motion — I. Numerische Math., 25, 323–346.

58. LaBudde, R. A., and D. Greenspan (1976). Energy and momentum conserving methods of arbitrary order for the numerical integration of equations of motion — II. Numerische Math., 26, 1–16.

59. LaBudde, R. A., and D. Greenspan (1978). Discrete mechanics for anisotropic potentials. Virginia J. Science, 29, 18–21.

60. Leith, C. E. (1967). Numerical hydrodynamics of the atmosphere. Proc. Symp. Appl. Math., Amer. Math. Soc., Providence, Rhode Island, 125–137.

61. Li-Shang, C. (1965). Existence and differentiability of the solution of a two-phase Stefan problem for quasilinear parabolic equations. Chinese Math., 7, 481–496.

62. Lorente, M. (1974). Bases for a discrete special relativity. Publ. #437, Center for Theoretical Physics, MIT, Cambridge, Mass.

63. MacPherson, A. K. (1971). The formulation of shock waves in a dense gas using a molecular dynamics type technique. J. Fluid Mech., 45, 601–621.

64. May, R. M. (1975). Biological populations obeying difference equations: stable points, stable cycles, and chaos. J. Theor. Biol., 51, 511–524.

65. Meyer, G. H. (1973). Multidimensional Stefan problems. SIAM J. Num. Anal., 10, 522–538.

66. Mills, R. D. (1965). Numerical solution of viscous flow equations for a class of closed flows. J. Roy. Aero. Soc., 69, 714–718.

67. Møller, C. (1972). The Theory of Relativity, 2nd Edition. Clarendon Press, Oxford.

68. Monin, A. S. (1972). Weather Forecasting as a Problem in Physics. MIT Press, Cambridge, Mass.

69. Morse, P. M., and H. Feshback (1953). Methods of Theoretical Physics. McGraw-Hill, New York.

70. Muetterties, E. L. (1977). Molecular metal clusters. Science, 196, 839–848.

71. Muirhead, M. (1973). The Special Theory of Relativity. Wiley & Sons, New York.

72. Newton, I. (1971). Mathematical Principles of Natural Philosophy. Univ. California Press, Berkeley, Calif.

73. Osterby, O. (1974). Analysis of numerical solution of the Stefan problem. DAIMI PB-24, Dept. Comp. Sci., Univ. Aarhus.

74. Pan, F., and A. Acrivos (1967). Steady flows in rectangular cavities. J. Fluid Mech., 28, 643-655.

75. Pasta, J. R., and S. Ulam (1959). Heuristic numerical work in some problems of hydrodynamics. MTAC, 13, 1-12.

76. Popov, Yu. P., and A. A. Samarskii (1970). Completely conservative difference schemes for the equations of gas dynamics in Euler's variables. USSR Comp. Math. and Math. Phys., 10, 265-273.

77. Potter, D. (1973). Computational Physics. Wiley & Sons, New York.

78. Preisendorfer, R. W. (1965). Radiative Transfer on Discrete Spaces. Pergamon Press, New York.

79. Remson, I., G. M. Hornberger, and F. J. Molz (1973). Numerical Methods in Subsurface Hydrology. Addison-Wesley, Reading, Mass.

80. Resnick, R. (1972). Basic Concepts in Relativity and Early Quantum Mechanics. Wiley & Sons, New York.

81. Richtmyer, R. D., and K. W. Morton (1967). Difference Methods for Initial-Value Problems, 2nd ed. Wiley & Sons, New York.

82. Saffman, P. G. (1968). Lectures on homogeneous turbulence. Topics in Non-linear Physics, 485-614. Springer-Verlag, New York.

83. Saltzer, C. (1964). Discrete potential and boundary value problems. Duke Math. J., 31, 299-320.

84. Schlichting, H. (1960). Boundary Layer Theory. McGraw-Hill, New York.

85. Schubert, A. B., and D. Greenspan (1972). Numerical studies of discrete vibrating strings. TR 158, Dept. Comp. Sci., U. Wisconsin, Madison.

86. Schulman, E. (Editor), (1970). Modern Developments in the Theory of the General Ocean Circulation — A Symposium. National Center for Atmospheric Sciences, Boulder, Colorado.

87. Schultz, D. H. (1973). Numerical solution for the flow of a fluid in a heated closed cavity. QJMAM, 26, 173-192.

88. Schwartz, H. M. (1968). Introduction to Special Relativity. McGraw-Hill, New York.

89. Soos, E. (1973). Discrete and continuous models of solids III. Stud. Cerc. Mat., 25, 687-759.

90. Stephani, L. M., and T. D. Butler (1975). A numerical method for studying the circulation patterns of a fluid in a cavity. Rpt. LA-6014. Los Alamos Sci. Lab., Los Alamos, New Mexico.

91. Streeter, W., and E. Wylie (1975). Fluid Mechanics. McGraw-Hill, New York.

92. Synge, J. L. (1965). Relativity: The Special Theory. North-Holland, Amsterdam.

93. Taylor, E. F., and J. A. Wheeler (1966). Spacetime Physics. Freeman, San Francisco.

94. von Karman, T. (1963). Aerodynamics. McGraw-Hill, New York.

95. von Neumann, J. (1963). Proposal and analysis of a new numerical method for the treatment of hydrodynamical shock problems. The Collected Works of John von Neumann, Vol. 6, No. 27. Pergamon Press, New York.

96. Whitrow, G. J. (1961). The Natural Philosophy of Time. Harper's, New York.

97. Wozniak, C. (1973). Foundations of the mechanics of discretized systems. Mech. Teoret. Stos., 11, 47-61.

98. Zabransky, F. (1968). Numerical Solutions for Fluid Flow in Rectangular Cavities. Ph.D. Thesis, Dept. Applied Math., Univ. Western Ontario, London, Canada.

99. Zeigler, B. P. (1976). Theory of Modelling and Simulation. Wiley & Sons, New York.

Index

Acceleration 2, 3, 9, 16, 18, 20,
    30, 46, 117, 125
  4-Acceleration 129
Angular
  momentum 27-30, 31
  velocity 65, 72, 73
Attraction 7, 18, 36, 64
Average velocity 58
Axiom of relativity 109, 117, 119,
    120, 126, 130

Backflow 92-104
Bar 34, 36
Building block 33, 49, 52

Center
  of gravity 25-27, 72, 74
  of mass 25-27, 72, 74
Charged particle 135
Chebyshev series 49
Clock 110, 136
Clusters 74
Collision 45
Computer 1, 3, 25, 39, 49, 55, 136
Conservation 4, 5, 12, 19, 22, 27-31
    44, 47
  of angular momentum 4, 27-30, 31
  of energy 4, 5, 12, 19, 22, 122,
    139
  of linear momentum 4, 27, 122, 131
  of momentum 4, 27-31, 119
Conservative models 11, 31-43
Contraction 111
Cross product 28, 30
Crystallization 48, 49

Damping 47, 52, 54, 92, 104
Density 44
Determinism 1, 109
Difference operator 116
Dilation 111
Dynamical equation 4, 21, 120, 125, 133

Earth 16, 69
Elastic
  collision 119
  vibration 36-39
Elasticity 36-39
Electric charge 145
Electromagnetic field 135
Electron 119, 121, 136
Energy 4, 5, 6, 11, 22, 121, 131
Equilibrium 52, 55, 64
Event 110
Explicit formulas 44-107

Finite element 49
Flow of heat 34
Fluid flow 39-43, 58, 71, 92-104
Force 4, 5, 7-19, 21, 30, 32, 49, 65, 139
FORTRAN 30, 65, 148, 154
Free
  boundary 48, 49
  surface 92-104

Galaxy arms 92
Gamma ray 121
Gas 44, 72
Generalized Newton's method 13-14, 24, 43
Gravitation 7-16, 20, 28, 64, 92

163